RESOURCE CONFLICT IN THE HORN OF AFRICA

🜲 PRIO

International Peace Research Institute, Oslo
Fuglehauggata 11, N-0260 Oslo, Norway
Telephone: (47) 22 54 77 00
Telefax: (47) 22 54 77 01
E-mail: info@prio.no

The International Peace Research Institute, Oslo (PRIO) is an independent international institute of peace and conflict research, founded in 1959. It is governed by an international Governing Board of seven individuals, and its main source of income is the Norwegian Research Council. The results of all PRIO research are available to the public.

PRIO's publications include the *Journal of Peace Research* (1964–) and *Security Dialogue* (formerly *Bulletin of Peace Proposals*) (1969–) and a series of books. Recent titles include:

Robert Bathurst: *Intelligence and the Mirror: On Creating an Enemy* (1993)

Nils Petter Gleditsch et al.: *The Wages of Peace: Disarmament in a Small Industrialized Economy* (1994)

Olav Schram Stokke & Ola Tunander, eds: *The Barents Region: Cooperation in Arctic Europe* (1994)

Kumar Rupesinghe & Khawar Mumtaz, eds: *Internal Conflicts in South Asia* (1996)

Jørn Gjelstad & Olav Njølstad, eds: *Nuclear Rivalry and International Order* (1996)

Johan Galtung: *Peace by Peaceful Means: Peace and Conflict, Development and Civilization* (1996)

Pavel K. Baev: *The Russian Army in a Time of Troubles: From the Taiga to the British Seas* (1996)

Valery Tishkov: *Ethnicity, Nationalism and Conflict in and after the Soviet Union: The Mind Aflame* (1997)

Ola Tunander, Pavel Baev & Victoria Ingrid Einagel, eds: *Geopolitics in Post-Wall Europe: Security, Territory and Identity* (1997)

RESOURCE CONFLICT IN THE HORN OF AFRICA

JOHN MARKAKIS

 PRIO

International Peace Research Institute, Oslo

SAGE Publications

London • Thousand Oaks • New Delhi

First published 1998

 SAGE Publications Ltd
6 Bonhill Street
London EC2A 4PU

SAGE Publications Inc
2455 Teller Road
Thousand Oaks, CA 91320

SAGE Publications India Pvt Ltd
32, M-Block Market
Greater Kailash – I
New Delhi 110 048

British Library Cataloguing in Publication data

A catalogue record for this book is available from the British Library.

ISBN 0 8039 8847 8

Library of Congress catalog card number 97-061787

Typeset by M Rules
Printed in Great Britain by The Cromwell Press Ltd,
Broughton Gifford, Melksham, Wiltshire

*To the peasants and pastoralists of the region.
Hardworking and knowledgeable, may they
overcome the obstacles nature and other men
set before them.*

Contents

List of Maps

Foreword

Since about the mid-1980s, there has been an increasing amount of scholarship on the relationship between issues of peace, conflict and security on the one hand, and environmental degradation and growing scarcity of natural resources on the other. This work has shone a light on some relatively overlooked aspects of the complex relationship between human society and its natural environment. Though the connections are often indirect, and though a great deal more research is required before the links are fully understood, it is now clear that the consequences of human impact on the environment can include social and political conflict. One of the ways in which human society adjusts in the face of environmental degradation is through conflict, often violent. It is thus important to integrate some of the concerns of peace research, development studies and scholarship on natural resources and the environment.

The Horn of Africa is a region that has suffered and continues to suffer from a double burden of severe environmental problems and extremes of violent conflict. Warfare in Ethiopia from 1962 to 1991 resulted in one and a half million deaths. Warfare in Sudan from 1955 until 1972 and again from 1983 until the present day has resulted in over one million deaths. War in Somalia in the 1990s destroyed the state and impoverished the people. In all three countries, warfare was accompanied by major famines: in Ethiopia from 1984 to the end of the war; in Sudan in 1988 and 1991; and in Somalia in 1992.

This study by Professor Markakis not only registers the severity of the human tragedies of the region and the depth of the social and environmental problems that underlie them; it also reveals the close links between the underlying problems. The people of the Horn are caught in a web of deprivation and powerlessness. The prevalent forms of state power for most of the past four decades have marginalized most people. The majority have also been unable to obtain adequate access to natural resources for food security and the other basics of a decent life. The result has been uprising and armed conflict. Though the wars are fought over a variety of issues and under various banners, the basis of the problems is the failure of development in the region and the pressure that puts on social groups to secure their position by attempting to take control of land and water resources.

This study has had a long genesis. The International Peace Research Institute, Oslo (PRIO) has conducted research on the relationship between peace and conflict issues and the environment since the late 1980s. For several years the institute collaborated with the United Nations Environment Programme (UNEP). In September 1991, as part of this continuing effort, PRIO and UNEP jointly convened a seminar in Sigtuna, Sweden, to discuss resources, conflicts and cooperation in the Horn of Africa. The intention behind the gathering was not only to have serious discussions between experts, but also to compile the seminar papers into an anthology. Those who have experience in putting together collections of papers presented at seminars and conferences know the pitfalls of such an enterprise. Unfortunately, although many of the papers were of very high quality, the resulting anthology was somewhat uneven, did not cover the full range of problems and contained internal inconsistencies (for example, experts disagree over the geographical definition of the Horn of Africa – what countries it includes – and this basic discrepancy of views was not reconciled within the collection of seminar papers). In early 1993, therefore, PRIO and UNEP decided not to publish the anthology but instead to attempt an alternative way of bringing the results of the seminar to a wider public.

We approached John Markakis of the University of Crete and asked if he would take on the task of compiling the book himself, utilizing the seminar papers or parts of them when possible, and rewriting, reconstructing and providing completely new material when necessary. With the passage of time since the original seminar papers were written and the unfolding of key events in the region – not least in Ethiopia and Somalia – it turned out that Professor Markakis had taken on not only a large drafting and editorial task, but also a major research job. As the work developed, it became clear that this was his book – not a compilation of others' work, but his own, while also drawing in some of the papers from 1991 seminar. In these places in the text that follows, the original contribution is acknowledged. I must thank both the seminar participants whose work has been used in this way and those whose work was not so used, for their generosity and forbearance.

It is also a pleasure to acknowledge and express PRIO's gratitude to the Norwegian Ministry of Foreign Affairs for its financial support of this work, and UNEP for both its financial and substantive role in developing this project over several years. Though the book's gestation has been long, the efforts put into it have been justified by the result, a multidimensional survey of what is arguably the world's most troubled region, with an analytic and diagnostic eye that also produces insights into more peaceful and fruitful possibilities.

Dan Smith
Director, PRIO
Oslo
February 1997

INTRODUCTION

Today we have come to realize that changes in the environment and the availability of natural resources have appreciable socio-economic impact, and these factors figure increasingly in political analysis. The attention of social scientists has focused particularly on the impact of environmental degradation and resource scarcity on social and political stability. There is growing conviction that correlations between them are discernible, and rising concern that 'environmental degradation and the wastage of renewable natural resources are playing important and growing roles in causing wars, regional and national insecurity, internal strife, and bloodshed' (Timberlake & Tinker, 1985, p. 59).

The authors of this statement noted that this malignant correlation was not 'widely documented' and there was little direct evidence of a causal relationship. Since then, social scientists have tried to document this link and to establish a causal relationship between environmental degradation, social unrest and political instability. Research has focused on case-studies concerning mainly the developing world, and numerous publications have appeared, some bearing explicit titles like 'Environment and Conflict' (Earthscan, 1984), *Global Resources and International Conflict. Environmental Factors in Strategic Policy and Action* (Westing, 1986), *Greenwar* (Bennett, 1991), *Les Conflits Verts* (Schmitz, 1992), *The Environment and International Security* (Lodgaard & Hjort af Ornäs, 1992), *Environment, Poverty, Conflict* (Graeger & Smith, 1994). These efforts have strengthened the conviction that discernible correlations exist between these unwholesome phenomena. According to a recent statement, 'environmental scarcities are already contributing to violent conflicts in many parts of the world', and 'these conflicts are probably the early signs of an upsurge of violence in the coming decades that will be induced or aggravated by scarcity' (Homer-Dixon, 1994, p. 6).

The accumulation of case-studies has yet to produce a conceptual breakthrough, and the precise definition of a causal relationship remains elusive. Advocates of the hypothesis that links environmental degradation to social conflict have sought to raise its profile and give it political clout by relating it to the security of nations and states. An early formulation of 'environmental security', sponsored by the United Nations Environment Programme, stated: 'Increasing levels of insecurity and instability within and among countries have stemmed from environmental stress and degradation, caused by disruption of the environment through armed conflict, excess pollution, and unsustainable exploitation of natural resources' (UNEP/PRIO, 1989, p. 9). In 1994, the World Bank added environmental security to its concerns (World Bank, 1994, p. 24), and US Secretary of State Warren Christopher declared that, 'as we move to the 21st century, the nexus between security and the environment will become even more apparent' (*Washington Post*, 15 April 1996).

   The notion 'environmental security' was linked to the concept of 'sustainable development', an innovation in economic theory brought to the world's attention by the Brundtland Report of the World Commission on Environment and Development (WCED) in 1987. Sustainability requires that the rate of consumption of renewable natural resources should not exceed their rate of renewal, and the disposal of wastes should not exceed the earth's capacity to absorb them. 'Indeed, without an inflexible commitment to the sustainable development of resources and the sustainable disposal of wastes there can be no environmental security' (Westing, 1991, p. 168). Since natural resource distribution follows eco-geographical rather than geopolitical lines, environmental security is thought unlikely to be achieved within the boundaries of single states. Consequently, an approach based on eco-regional cooperation is recommended (Westing, 1989). Before long, the concept of sustainability acquired global dimensions, which were much emphasized in the United Nations Conference on Environment and Development (UNCED), held in Rio de Janeiro in 1992. The 'Earth Summit', as it was dubbed, placed the globe's resources under universal stewardship, challenging the moral, if not the legal, right of societies and states to dispose of them as they see fit.

   The concept of sustainability has its critics. They find the coupling of sustainability and development incongruous, and wonder what level of consumption is globally sustainable. If, as commonly perceived, the standards of development are set by the industrialized societies, the concept is a hoax, for the earth obviously cannot sustain all its inhabitants at such levels. The critics charge that the concept is designed to absolve developmentalism – 'that race without a finishing line' (Sachs, 1993, p. 8) – of having sinned against nature by giving it an environmental consciousness; the objective being to conserve development, not to conserve nature. Globalization, they allege, means that the management of the earth's resources becomes a matter of world politics, where the powerful industrial nations hold sway and can defend the exceedingly high consumption levels they attained by exploiting and depleting the resources of the globe.

   The notion of environmental security also has its critics (see, e.g., Dokken & Graeger, 1995; Deudney & Mathew, forthcoming). Some of them think it is a mistake to raise environmental issues to the level of traditional politico-military security concerns, because it might lead to the militarization of environmental issues. Others question the basic premise linking environmental degradation to conflict (Painchaud, 1990). Carried out among scholars in the West, the reference point of the debate is the developed world and its concern for global stability. The type of conflict that concerns the participants is conventional warfare between states; one of them states categorically that 'environmental degradation is not very likely to cause interstate wars' (Deudney, 1990, p. 461). Inconsistently, however, the empirical content of the debate is drawn almost entirely from the developing world, where the predominant form of conflict is civil war (Wallensteen & Axell, 1993). In the perspective of the developed world, violent societal conflict caused by resource scarcity, itself the result of environmental degradation, seems a remote prospect. It is assumed that human activity harmful to the environment can be halted without great cost to society. It is further assumed that resource scarcity can be relieved by trade, or overcome by technological innovation (Deudney, 1990, pp. 470–471). Obviously, these assumptions do not apply to most of the developing world.

There is little doubt that environmental degradation does lead to resource scarcity, especially of land, water and woodfuel in many parts of the developing world. This is a process that can neither be halted at present, nor relieved by technological innovation or trade. It is also evident that unrelieved scarcity intensifies group competition over diminishing resources, and this can lead to social conflict. Whether it does or not depends largely on the context – economic, social, political – within which competition is waged. In the developed world, group competition for resources is waged through institutionalized economic and political processes, sanctioned by culture, defined by law and upheld by the state. Social conflict occurs only when these processes break down under intolerable pressure. The improbability of such a breakdown is one of the blessings of economic and political development. It is the absence of institutionalized control in global competition that makes interstate war the threat that looms largest in the developed world's perspective of security.

By contrast, in most developing countries, institutionalized processes regulating group competition over resources have merely a formal existence. Raw political power utilizing the instrumentality of the state is the regulating factor in this competition. Consequently, the role of the state is a key variable in the process of conflict generation. In many developing countries, especially in Africa, the state controls the production and distribution of material and social resources. Access to state power ensures access to resources, but not all groups have equal access, and some have none at all. Those that do, use the power of the state to protect their privileged position; those that do not, struggle for state power in order to gain access to resources. Thus, in many countries of the developing world, the state is both the bone of contention and a major actor in social struggles over resources – a far cry from the developed world's perception of the state's proper role.

The perception of the state as a neutral arbiter – a well-tended liberal myth in the West – also depreciates its role in the rest of the world. The state's involvement in the production and distribution of material and social resources in many developing countries does not only determine the pattern of wealth distribution and access to facilities. It also helps to determine the pattern of resource use which, in turn, affects the environment. For example, state promotion of commercial agriculture greatly limits the size and quality of land available for subsistence farming. This has adverse environmental consequences in both sectors. Intensive commercial agriculture strips the land of its natural cover and exhausts the soil, while subsistence cultivators are crowded into areas of inadequate size and inferior quality, with similar results for the environment. From this vantage point, environmental degradation is the result not only of people's interaction with their habitat but also of social and political patterns determined and maintained by state power.

The influence of material factors in social interaction, particularly in the generation of conflict, is similarly depreciated in the industrialized world, where social science analysis emphasizes complexity in human motivation and multi-causality in behaviour. As a result, the primacy and immediacy of the material factor in subsistence economies is underestimated. In most parts of the developing world, the condition of the local environment is crucial to the ability of societies to sustain themselves. Subsistence cultivation uses no inputs to mediate its productive interaction with the environment; consequently, it is entirely dependent upon it. Subsistence economies have no surplus, no reserves, little capacity for adaptation through technological innovation, and little or no state support. In these circumstances, environmental change with a negative effect on production, i.e. environmental degradation, has an early and dramatic impact on people's lives.

Various forces – demographic, climatological, economic, political – combine to launch a destructive assault on the environment. This poses a clear threat to the ability of many societies in the developing world to sustain themselves. It is extremely difficult to halt this process at present since the very lives of people depend on it: 'People who are dependent on nature for their survival have no choice other than to pursue the last remaining fragment of its bounty' (Sachs, 1993, p. 10). For example, deforestation cannot be stopped as long as the need for fuel and additional land for cultivation cannot be met otherwise. In this way, environmental degradation exacerbates resource scarcity and intensifies competition. Intensified resource competition on the one hand, and the absence of effective regulatory processes on the other, combine to make social conflict more likely in some parts of the world.

The various ways in which resource scarcity and competition can fuel conflict are not obvious in most instances, nor is the connection necessarily direct. Often, the outsider's perception of the conflict is not related to its objective causes. Even more confusing, the parties to the conflict themselves may prefer less mundane causal explanations and resort to ideological expressions involving ethnicity, religion, nationalism, etc. A fairly common instance can illustrate the complexity involved. People often drift away from areas where the land is losing its productivity because it is either overpopulated, overworked, or degraded, and move into less densely populated areas where productive land is available. Over the years, the new area may become congested, and competition over land may develop, leading to conflict. One form such conflict can take is to pit the original inhabitants against the immigrants. If these two groups claim different ethnic identities, the ethnic factor will easily overshadow all others, and the conflict is likely to be perceived by outsiders, even by many of the participants, as ethnic conflict.

An incident of this kind occurred in Kenya in 1993, when Kikuyu peasants who had moved into the Rift Valley in large numbers since the 1960s came under violent attack by Kalenjin tribesmen who claim prior rights of possession to the region. The Kikuyu move took place under the Kenyatta regime which was dominated by their ethnic group. The Kalenjin attack occurred when Arap Moi, an ethnic kinsman, ruled Kenya. Although the importance of the land factor was generally recognized, the clashes were widely attributed to tribal enmity inflamed by political passion.

The Horn of Africa is a textbook case of environmental degradation and conflict. (For a bibliography see IUCN, 1991.) The region has suffered monumental environmental damage, the work of natural forces and human activity, seriously impairing the capacity of many of its inhabitants to sustain themselves. With the exception of Kenya, the other states of the region – Ethiopia, Eritrea, Sudan, Somalia, Djibouti – are among the poorest in Africa, and famine is a frequent visitor there. Manifold social conflict is an endemic feature in the recent history of the Horn of Africa. Such conflict has deep roots in the history of the region and a plethora of discernible causes. It is fought under many banners, representing nations, ethnic groups, regions, religions, social classes and clans. Historical, cultural and ideological factors certainly play a role, but material factors are by far the most prominent, and land is the essence. One of the most significant, yet little noted, consequences of conflict in many parts of the Horn is the appropriation of land by groups that gain local advantage through association with the state or by naked force of arms. Numerous instances of this are cited in this study. Water is another scarce resource in this region and the object of contention among groups, as well as a potential source of inter-state conflict.

The analysis of conflict offered in this study indicates a clear correlation between exclusion from state power, reduced access to resources and the incidence of conflict. Simply put, areas whose natural endowment is poor and depleted, and where little or no compensating development has taken place, are also areas where conflict has flourished. Despite tenacious efforts by ruling groups in the Horn over four decades, with the generous assistance of the world's greatest powers, it proved impossible to contain the conflict militarily. What has been achieved instead is the disruption and dislocation of fragile economies, loss of food security and the reduction of the region's population to destitution and periodic famine.

The Horn is not merely a geographic designation. It is a region with a multitude of physical, social, cultural, economic and other integrating features that distinguish it from adjacent regions. Its sea boundaries are the Red Sea and the Indian Ocean. Its inland boundaries are two prominent landmarks – the East African highlands and the Nile basin – which also represent social and cultural watersheds separating the people of the Horn from those of adjacent regions. Intercourse among the peoples of the Horn has been continuous since early times. Extensive migrations within it brought into intimate contact, even mingled, its main population groups. Many have mixed origins, speak related languages, and have close cultural ties. The Horn is also what is termed an eco-geographical region: one that is unified in an ecological sense, and is distinct in that sense from the regions contiguous to it. Ecological zones within the region span the territories of the several states, as do the material cultures that evolved in these zones and the societies that produced them.

   Accordingly, the perspective adopted by this study is regional. The aim is
to depict the common and universal features, the continuities, connections
and interactions that have shaped the destiny of the peoples of the region, cutting across what are proving to be ephemeral politico-administrative
boundaries. The fragility of the geopolitical pattern in the Horn was demonstrated in the beginning of the 1990s with the separation of Eritrea from
Ethiopia and of Somaliland from Somalia. The regional perspective fits as
well the pattern of conflict in the region, which is seldom confined within the
borders of one state. Finally, in view of the lack of correspondence between
ecological zones and material cultures on the one hand, and politico-administrative boundaries on the other, some of the major environmental issues in
the region are not approachable at state level. A regional approach is
demanded by the nature of the issues themselves.

AUTHOR'S NOTE
The validity of statistical data pertaining to the countries in the Horn of
Africa is questionable. They should be read only as indicators of the level of
magnitude, not as precise figures. Often, figures cited in different sources do
not agree.

PART ONE
POPULATION, RESOURCES, FOOD SECURITY

1

Population

1.1 Current Trends

People, it would seem, are the most perishable resource in the Horn of Africa. In recent years an enormous toll in lives has been taken by war, famine and disease, as the Four Horsemen of the Apocalypse criss-crossed the region, leaving few communities unscathed. Countless people were physically and mentally maimed and left to live handicapped lives. Vast numbers of people have been stampeded across international borders, herded into refugee camps, packed into famine relief shelters, or crowded into peri-urban slums. The image of a cadaverous child holding forth an empty bowl has become the emblem of the Horn of Africa. Far more impressive than any statistic, it has sealed the impression the world has of the Horn as a region with more people than it can feed. This chapter will examine population trends in the Horn, and later chapters will relate them to the region's resources.

Certainly, the Horn does not seem a crowded region (see Table 1.1 and Table 2.1 below). Sudan, Africa's largest country in area, has a population of about 25 million and the lowest overall density. With half the area of Sudan, Ethiopia has twice as many people and probably the highest overall density. Somalia is a little over half the size of Ethiopia, and has perhaps one-sixth of the latter's population. With only a little less territory than Somalia, Kenya has perhaps three times as many people. Eritrea occupies an area of 125,000 square kilometres and claims a population estimated at 3.2 million. The Djibouti enclave measures 23,000 square kilometres, and has a population of about half a million.

The countries of the Horn have high reported rates of population growth. With the apparent exception of Djibouti, they range between 3 and 4% annually (see Table 1.1). The region follows the general trend in sub-Saharan Africa, where the average for the decade of 1980–1990 was 3.07%. Population growth is due to a downward trend in mortality rates, while fertility rates have remained consistently high. The highest fertility rate ever for any country was recorded in Kenya in the 1970 census: an average of 7.9 births per woman (see Kenya, 1989, p. 207). The rate for Ethiopia in the early 1990s was estimated at 7.7

Table 1.1 *Population*

Country	Population last census	Growth rate (%)	Population estimate 1994	Population projection 2050	Population under 15 years (%)
Djibouti	125,000 1973	2.5	500,000	1,850,000	36.9
Eritrea	2,621,566 1984	–	3,200,000	–	–
Ethiopia	42,616,876 1984	3.0	54,000,000	148,179,000	48.9
Kenya	21,443,636 1989	3.9	25,400,000	100,000,000	49.3
Somalia	7,114,431 1987	3.1	8,954,000	23,944,000	46.0
Sudan[1]	21,267,000 1993	3.2	28,129,000	82,753,000	46.0

[1] Population and growth rate cover only the North.

Sources: Sudan, 1995. *Fourth Population Census of Sudan 1993. Northern States.* Department of Statistics, Khartoum; UNESCO, 1995. *Demographic Yearbook 1993*; UNESCO, 1995. *Population and Vital Statistics, mid-year 1994 estimates*; UNDP, 1995. *Human Development Report, 1995*, New York.

(see Ethiopian Transitional Government, 1993, p. 28). The rate for Northern Sudan during the same period was thought 'likely to be 7.5 or more children per woman' (Sudan, 1995, p. xii). These rates are higher than the average rate for Africa, which was 6.4 in the 1980s. The decrease in mortality rates is due mainly to significant reductions in infant and child mortality thanks to medical intervention; in itself a considerable achievement. Life expectancy has also increased. Kenya ranks highest in the region – 55 years – while Ethiopia, Eritrea, Somalia and Djibouti have life expectancy rates below 50 years (see Table 1.2).

Ironically, having fuelled the so-called population 'explosion', these achievements are widely assumed to have undermined the prospects of socio-economic development in these countries. This assumption is based on projections of present trends. Kenya's population increased five-fold between 1948 – when the first census was taken – and 1990. The other countries doubled their population in the same four decades, and are projected to grow prodigiously in the future (see Table 1.1). Ethiopia, whose population was estimated at less than 20 million in 1950, is expected to approach 150 million by the year 2050. Kenya, at 5.2 million in 1948, is projected to reach 100 million in 2055. Besides sheer size, the population of the Horn has a very large proportion – 47–48% – of its members under 15 years of age (the average for Africa is 45%). In addition to the heavy burden it imposes on material and social resources, this demographic factor ensures population will continue to grow even when fertility rates fall, because of the large proportion of people in the younger age groups. Although women may bear fewer children, there are more women and men entering the reproductive stage of life.

Table 1.2 *Social and Economic Indicators*

Country	Urbanization (%)	GNP per capita ($)	Fertility rate (1993)	Infant mortality per thousand	Life expectancy (1992)
Djibouti	75.0	780 (1993)	5.8	114.9	48.3
Eritrea	15.1	–	–	105.2	48.0
Ethiopia	15.0	100 (1993)	7.7	119.2	47.5
Kenya	20.6	270 (1993)	6.3	69.3	55.7
Somalia	23.5	120 (1990)	7.0	121.7	47.0
Sudan	26.5	70 (1985)	7.5[1]	78.1	53.0

[1] *Fourth Population Census of Sudan 1993* gives a fertility rate of 6.2% for Northern Sudan.

Sources: UNDP, 1995. *Human Development Report, 1995*. New York; Sudan, 1995. *Fourth Population Census of Sudan 1993. Northern States*. Department of Statistics, Khartoum; *National Population Policy of Ethiopia*, Addis Ababa, 1993; UNEP, 1993. *Environmental Data Report, 1991–1992*, Nairobi.

The linkage between population growth and development has been the subject of intense debate for some time (see Cassens, 1994). The first United Nations World Population Conference (1974, Bucharest) focused on the need to understand this linkage. It adopted a World Population Plan of Action, to 'advance national and international understanding of the complex relations among the problems of population, resources, environment and development'. In 1977, the United Nations established an Interagency Task Force on Inter-Relationships between Population and Development. The United Nations Fund for Population Activities sponsored research and financed a series of conferences and publications on this subject. The 1984 International Population Conference (Mexico City) urged governments to adopt policies, including population control measures, designed to redress the balance between population, resources and the environment. An alarmist note flavoured the debate, as shown in the use of terms like 'the population bomb' and 'population explosion' (Ehrlich, 1990).

Population growth and development thus became intrinsically linked, although there is no agreement on the precise nature of this link (Turner et al., 1993). Reflecting Western expert opinion, most international agencies regard Third World population growth as an obstacle to development. The report of the World Commission on Environment and Development (WCED, 1987) argued that environmentally sound and sustainable development can be attained only if population size and growth rates are commensurate with the productive potential of nature. The International Forum on Population for the 21st Century (1989, Amsterdam) issued a 'Declaration on a Better Life for Future Generations', calling for a commitment to a sustainable relationship between population numbers, natural resources and development. Finally, the agenda for the 21st century (the so-called *Agenda 21*) adopted at the United Nations Conference of Environment and Development (1992, Rio de Janeiro) is replete with recommendations for control of demographic trends.

These are based on the assumption, spelled out in the report of the United Nations Environment Programme, *The World Environment, 1972–1992* (UNEP, 1992b), that a continued rapid increase of the world's population will seriously affect the possibility of achieving widespread improvement in the quality of human life and sound management of natural resources. In 1993, national academies of science in 58 countries joined in a call to governments to take action on population growth, based on their concern that a higher quality of life is not possible for all people on the globe given the current trends in population growth and resource use (National Academy of Science, 1994).

A Malthusian implication is evident in the assumption that regards population growth as an obstacle to development. It is linked to the notion of the so-called 'carrying capacity' of agricultural land under existing modes of production. The implication is that population size cannot exceed the carrying capacity of land, which, in turn, is determined by the existing technology of production. Whereas overall population density in the Horn is low, the situation is dramatically different when cultivable land is considered (see Table 2.1). For example, overall population density in Sudan was 11 persons per square kilometre in 1993, while the figure for arable and irrigated land was more than ten times that. In this region, the technology of subsistence agricultural and pastoralist production has changed little since pre-colonial times, and the annual rate of increase in food production – seldom more than 2% – has consistently fallen behind the rate of population growth. Furthermore, there is a multiplier effect involved in the deleterious impact of large numbers of people on the environment, which degrades the natural resource base upon which production depends, further diminishing the carrying capacity of land.

The situation in the Horn, and in Africa generally, is contrasted to the experience of the advanced countries in the West during a comparable period of their development. Europe's population increased in the last century, but the rate of growth was seldom more than 1.5%, because fertility rates declined along with mortality, producing what one study called a 'quiet revolution' (Gillis et al., 1992). More to the point, Europe experienced a technological revolution, and production rose faster than population to provide food and employment for its people. In the European periphery, where technological advance was slower, emigration to the New World provided a safety valve. For these reasons, even the iconoclastic agricultural economist René Dumont was moved to declare 'the real problem facing black Africa as we near the end of the 20th century is the population explosion' (*Le Monde Diplomatique,* May, 1990).

The Malthusian perspective with its dismal prospects is subject to many challenges. To begin with, its statistical integrity is open to doubt, given the often dubious nature of the data upon which it is based. In the Horn, only Kenya offers population censuses that are considered reliable. Somalia has never had a census, and existing estimates of its population range between 5 and 8.5 million. The latter figure was interpolated by the Somali government of the time from the returns of an election held in 1986. Since it represented a 30% increase on the highest previous estimate, it was considered dubious by the international agencies; each of them in turn adopted its own estimate, with total confusion as the result (ILO, 1989, pp. 163–164). Ethiopia's first attempt to count its population was made in 1984, but this was only a partial success because it did not cover several war-torn regions. Even so, it is supposed to have revealed a growth rate much higher than previously assumed, and the country's overall population estimate jumped by several millions. Another census was carried out ten years later, the results of which were still awaited in 1996. Independent Eritrea's initial population estimate of 3.2 million was estimated from the votes cast in the independence referendum of 1993. This figure was challenged in subsequent negotiations with international aid agencies and was unofficially scaled down by several hundred thousand. Population statistics in Djibouti are subject to numerous caveats. The flow and ebb of waves of refugees, the rootlessness of pastoralists, and the political rivalry between the Afar and Somali inhabitants make population counts highly controversial exercises – so much so, in fact, that the results of a census taken in 1972–73 did not appear until four years later, and then only in fragmentary fashion. (The exercise was not repeated.) Population projections generally fail to take into account the impact of widespread and protracted war, persistent food deficiency, periodic famine and massive population displacement – all familiar features of recent history in the Horn. For example, Sudanese censuses fail to reflect population fluctuations in Southern Sudan, where, according to a study commissioned by the United States Congress, at least 1.3 million people lost their lives in the preceding decade (Burr, 1993). Sudan's latest census in 1993 excluded the Southern region altogether, and ingenuously recommended 'caution in using the results' (Sudan, 1995, p. xi). Nor do estimations of Somalia's population take into account child mortality rates, such as those reported by UNICEF (1996); according to this report, half or more of all children under the age of 5 on 1 January 1992 were dead at the end of the year.

'One reason why so many Ethiopians are so poor is that there are so many of them' (Pickett, 1991, p. 173): this negative correlation of rapid population growth to development is challenged by those who regard low, rather than high, population density as an obstacle to development. Their argument is based on the economies of scale and the size of the market. They point out that population density in the developed world is far higher than nearly everywhere in Africa, especially the Horn. This is true for arable land as well – the world average is 301 per square kilometre. Additionally, there are those who agree with Ester Boserup (1981, 1993), who sought to reverse the Malthusian equation by arguing that population pressure is the catalyst that can lead to changes in the mode of agricultural production and to higher productivity. In this perspective, population growth can be the catalyst for technological and social change. The case of the Machakos district in Kenya is often cited in support. Suffering from severe land degradation as early as the 1930s, the district was considered overpopulated in 1948, when the average density was 67 persons per square kilometre. By the early 1990s, density in some parts of the district exceeded 400, yet the value of output per hectare and person had increased tenfold and threefold respectively, and land degradation had been halted. Innovation, investment, diversification, transportation, commercialization, combined with tree planting and conservation measures, accounted for this success story (Tiffen et al., 1994). Finally, it is asserted that people are not poor because they have many children. Rather, they have many children because they are poor. Poverty raises the value of children as workers who contribute to the survival of the family (Mamdani, 1972).

The debate on population growth is charged with political interest and ideological passion. Third World ideologues suspect the interest of the developed world in population control derives more from a desire to retain the lion's share of world resources, rather than to promote Third World development. The Roman Catholic Church opposes population planning on doctrinal grounds, and other militant groups in the West oppose it on moral grounds. Western ideologues were joined by prominent Islamic religious leaders in opposition to the 1994 World Population Conference sponsored by the United Nations (Cairo). The Conference itself was sidetracked into debating the morality of abortion, homosexuality and family relationships, none of them major Third World concerns. However, the programme of action it adopted addressed some of the socio-economic and political issues related to the status of women and population growth (United Nations International Conference on Population and Development, 1995).

Whether rapid population growth is a bane or a boon to development, there is very little decision-makers in the Horn can do about it. Population control programmes relying on persuasion had little success world-wide, probably because they seek to lower fertility as a prerequisite to development, while the evidence of history suggests fertility declines only when development has reached an advanced stage. In other words, development is the best contraceptive. The Sudan Demographic and Health Survey 1989/1990 (Sudan, 1990) reported that women with secondary education in Northern Sudan had an average of 3.3 children, as compared with 5.9 for women with no education at all. Furthermore, such programmes focus on techniques of contraception, ignoring the socio-political context that militates against this practice (Moore & Schurman, 1988). Only in China, where coercion is used, has there been a modest measure of success. The states of the Horn lack the coercive capacity for such a task and, at any rate, none has shown real concern over population growth. According to the prevailing attitude, the real problem is not population growth, but the failure of the economy to expand. The issue is not one of population control, but of economic management. 'It is neither the absolute level of population nor its rate of growth per se which is the major concern since with a better technological base and resource endowment Kenya could support a much larger population' (Kenya, 1989, p. 209). None of the states in the Horn has adopted a meaningful policy of population control. Nevertheless, the constraints of the Structural Adjustment Programmes under which they operate oblige their governments to pay lip service to the principle. For example, the document cited above goes on to say the Kenyan government 'is determined to control by all legitimate and acceptable means the size of the population to match available resources' (p. 210). The regime that came to power in Ethiopia in 1991 wasted no time in formulating a National Population Policy pledging to reduce the current fertility rate of 7.7 to 4 by the year 2015 (Ethiopian Transitional Government, 1993, p. 28).

Current population trends in the region are likely to persist. Should peace return to the Horn, population growth may even accelerate. On the other hand, the technology of subsistence production is unlikely to change in the near future, and population pressure on available resources will increase. The deleterious impact on the environment will also intensify and will likely affect the natural resource base. Even if population growth holds the promise of development for the future, for the present it can only mean stagnant or lower standards of living for the majority. In short, there is a real problem involved in rapid population growth. 'When a population is growing faster than food production and overall standards of living are falling one cannot pretend that a population problem does not exist by referring to unused resources and agricultural potential' (El Tay et al., 1988).

1.2 Migration

If the past is any guide, population growth would lead to increased movement of peoples within and across state borders. As we will see, such movement is an integral part of the regional political economy, and is itself often the cause of conflict.

The history of the Horn is marked by massive population movements, in the course of which the region's ethnographic profile has repeatedly been redrawn. All the major population groups have been involved. The Arabs came from Upper Egypt to Northern Sudan, following the Nile south to the 10th parallel North, where they were stopped by the immense swamp they aptly named the Sudd ('barrier') and thus failed to enter the area now known as Southern Sudan. They were predominantly nomad pastoralists, and to this day 'Arab' means just that in Sudan. The Somali, another nation of nomad pastoralists, embarked on an inland migration from the tip of the Horn which lasted for centuries and was not halted until very recently. Moving southwest along the valleys of the Shebeli and Juba Rivers and along the Indian Ocean littoral, they came within sight of the East African highlands. The Oromo, the largest nation in the Horn, were also mobile pastoralists when they moved in a wide front onto the Ethiopian plateau from the south and east. When their historic movement ended, the Oromo vanguard had reached the Blue Nile gorge in the west and had penetrated deeply along the eastern edge of the Ethiopian highlands as far north as Tigray Province. The Oromo rearguard stayed in the lowlands of Kenya as far as the Indian Ocean.

While such massive movements took place over the course of centuries, smaller shifts are known to have occurred more recently. The Beja pastoralists of eastern Sudan shifted southwards, towards better watered land in the Gash and Barka inland river deltas and the northern foothills of the Ethiopian plateau. One of their sections, the Beni Amer, moved into what was to become western Eritrea. The Issa Somali have been moving steadily northward into the Danakil. It was in the 19th century that the Rashaida, a small camel-raising tribe from Arabia, crossed the Red Sea into what became eastern Eritrea. It was at the close of that century, in the wake of the devastation caused by the 'triple disaster' of rinderpest, bovine pneumonia and smallpox in the 1880s and 1890s, that pastoralist groups in southern Ethiopia and northern Kenya, like the Borana, Gabbra, Rendille, Turkana and others, moved to the lands they now occupy.

Mobility is prevalent among pastoralists, but is not unknown among cultivators in the Horn. The Abyssinian shift southwards over the centuries is the most striking instance. From the northern part of the plateau, with Axum at its core, the centre of the Abyssinian state shifted steadily southwards over the centuries until, at the end of the last century, it came to rest at the very centre of the plateau, where Addis Ababa now stands. The Abyssinian expansion continued in the early part of this century into the homelands of the Afar and Somali in the east, the Oromo in the east, south and west, the Sidama in the south, the Anuak in the west, and along the border with Sudan, until it established the boundaries of what came to be known as the Ethiopian Empire.

Although massive population movements in the Horn have come to a halt, they are not completely exhausted. The Abyssinian drift from the eroded northern half of the plateau to the southern half continues to this day. A major portion of the land conquered by the Abyssinians in the southern plateau was expropriated by the state, and throughout this century northerners settled there assisted by land grants. The most recent wave of this movement was the resettlement programme carried out by the Ethiopian state in the 1980s. In that massive operation, over half a million people were moved from the provinces of Tigray, Wolo and Shoa, mostly to the south.

The introduction of state borders during the colonial period proved a restraint to migration, though not quite a barrier. Pastoralists ignored them as a matter of course. The Somali push westwards into northern Kenya had to be halted forcefully during the colonial period, by drawing a line inside the Northern Frontier District which the Somali were not allowed to cross. In the postwar period, the Ethiopian authorities vainly tried to keep the Somali east of the Genale Dorya River and out of Sidamo Province.

Population movements have a variety of causes (Parnwell, 1993). Undoubtedly, one of them is ecological imbalance caused by natural, human and animal activity. In the absence of reliable data, we can only speculate about the causes of earlier mass migrations in the Horn. One plausible explanation is that, in a region noted for its meagre and erratic rainfall, the pastoralists were seeking greener pastures and permanent water sources. Of course, pastoralists move also to escape enemies or livestock disease. The oral tradition of most pastoralist groups in the Horn is a long record of movement in search of pastures, water and safety for themselves and their animals. The Abyssinian shift away from the eroded lands of the northern plateau is clearly related to the degradation of the environment, which is the result not only of natural erosion, but excessive human and animal population pressure as well.

Population shifts have been the cause of endless strife in the Horn, setting the scene for many of the ongoing confrontations. The Abyssinian expropriation of land in the southern plateau established an antagonistic relationship between the empire builders and their subjects in that area. This had momentous political repercussions in Ethiopia, which have still to be finally resolved. The Somali move westwards brought them into conflict with the Borana and Gabbra Oromo in Kenya and southern Ethiopia. It also created hostility between the Somali and the governments of colonial and independent Kenya, and the area they inhabit in that country has never been free of conflict this century. The politics of Djibouti are confounded by the enmity between the Afar and the Issa Somali, caused by the steady incursion of the latter into the land of the former. On the border between north and south Sudan, Baqqara designs on Dinka land have spawned a vicious conflict that has become part of the civil war in that country. On the border between Sudan and Ethiopia, Nuer incursion into Anuak territory has resulted in a permanent state of war between these two groups. On the Ethiopia–Somaliland border, Ishaq intrusion into the Haud area of the Ogaden has been the cause for the oldest standing internecine Somali conflict.

The ecological factor in population movement and the resultant conflict can be observed, in a microcosm as it were, at present in the south Omo valley of Ethiopia. In that remote region, where the first sign of modernity appeared recently in the form of the Kalashnikov rifle, small pastoralist groups like the Dassanetch, Mursi, Nyangatom and Bodi are caught in a vicious cycle of escalating warfare, caused partly by the need to shift their habitat in order to adjust to changing ecological and demographic conditions. Anthropologists have recorded in detail the movement and strife that follow drought, changes in the flood of the Omo River, changes in population size, as well as the fearful impact of modern weaponry (Markakis, 1993).

Conflicts of the type mentioned above, which have their origin in group competition for resources, are often woven into larger political confrontations that involve other groups and motives as well, so that their distinctiveness and origins become obscured. Alienated from the modern state, pastoralists in particular have a tendency to join movements that challenge state authority. Nevertheless, the antagonists seldom lose sight of their own goals, and seek to exploit any advantage the larger confrontation may afford them to pursue their local struggle. Thus, two groups in a local confrontation are likely to join opposed sides in a larger conflict. If one joins the rebel side, the other is likely to support the state, and the weapons they obtain through these alliances will be used against each other.

1.3 Refugees and Displaced Persons

Refugees and displaced persons are involuntary migrants. Officially, a refugee is defined as someone who fled his country because of a 'well founded fear of persecution by reason of his race, religion, nationality or political opinion' (UNHCR Statute, Art. 6B). Since the end of World War II, Africa has seen numerous waves of such refugees. Some of them were the result of struggles for independence in Algeria, Angola, Guinea-Bissau, Mozambique and Zimbabwe. Others were the result of violent civil conflict within the post-colonial state in Nigeria, Burundi, Chad, Rwanda, Uganda, Congo (formerly Zaire) and the countries of the Horn. In recent years, refugees and displaced persons in the Horn constituted a massive population movement, the largest in the modern history of Africa. The United Nations Special Emergency Programme for the Horn of Africa (SEPHA) estimated there were nearly 9 million refugees and displaced people in the region in 1991 (UN SEPHA, 1991, p. 12).

Internal political conflicts in Ethiopia, Somalia and Sudan have produced successive waves of refugees since the early 1960s. These states are both the sources and the hosts of each other's displaced humanity, as most refugees in the Horn lack the resources to do more than to cross the nearest border into a neighbouring state. Relatively few manage to go further abroad, and these are usually townsmen with education and skills. The refugee camps of the Horn are filled with peasants and pastoralists, whose involvement in political conflict is often peripheral, but whose livelihood has been destroyed and their lives endangered by such conflict.

The first refugee wave originated in Eritrea in 1967, and was the result of the first pacification campaign launched by the Ethiopian Army against the Eritrean Liberation Front (ELF). The violence visited on the rural communities of western Eritrea at that time forced about 30,000 people, mostly pastoralists, to seek refuge in eastern Sudan. By that time, Sudan was producing its own wave of refugees, the result of the civil war in Southern Sudan. Sudanese Army operations forced many people from Bahr el Ghazal and Equatoria Provinces to flee across the border to Congo (Zaire) and Uganda, while others from Upper Nile Province crossed the border into Ethiopia. In mid-1967, the governments of Sudan and Ethiopia agreed to conclude a treaty on refugees, offering them the option of repatriation, or removal from the frontier to a resettlement site.

From that time onwards, many agreements were signed by these two neighbouring states. Their real purpose was not to help the refugees, but to end support for each other's rebels. None of these agreements was honoured, and Sudan became a haven for the Eritrean nationalists, while Ethiopia gave sanctuary to the rebels from Southern Sudan. The number of Eritrean refugees in Sudan rose gradually until 1974, when the change of regime in Addis Ababa escalated the conflict, forcing hundreds of thousands of Eritreans to leave their homeland. Most of them found refuge in Sudan. In the 1980s, they were joined there by people from Tigray Province, where a revolt was in progress led by the Tigray Peoples Liberation Front (TPLF), and others from the borderlands of western Ethiopia, where the Oromo Liberation Front (OLF) was active. Young, educated people from the towns of Ethiopia also fled to Sudan to avoid conscription in the armies the Ethiopian regime was raising to fight multiple internal wars.

Southern Sudan enjoyed a respite from civil war for a decade following the conclusion of the Addis Ababa Agreement in 1972, and the Southern Sudanese refugees were able to return to their homeland at that time. War began again in the early 1980s, when the Sudanese Peoples Liberation Army (SPLA) resumed the struggle against Northern domination. Refugees once again poured across borders, a large number settling in Ethiopia, where the SPLA was welcome as a retribution for Sudan's tolerance of the Eritrean, Tigrayan, Oromo and other Ethiopian rebels. In 1991, when another change of regime brought the TPLF to power in Addis Ababa, the SPLA lost its welcome and was forced to leave Ethiopia in a hurry. Many Southern Sudanese refugees in Ethiopia were then obliged to return to Southern Sudan in the midst of war and famine.

Another flow of refugees was generated by the conflict between Ethiopia and Somalia over the Ogaden and adjacent regions, the sole instance of inter-state conflict in the Horn in the post-colonial era. The 1963 uprising in the Ogaden produced the first wave of Somali refugees from Ethiopia, who fled to the Somali Republic. In the mid-1970s, guerrilla fighting resumed in the Ogaden and spread to Bale and southern Sidamo Provinces. It was followed in 1977 with an invasion and occupation of these regions by the Somali army. The Ethiopian counter-attack that expelled the Somali forces from the area the following year caused a wholesale exodus of Somali and Arsi Oromo refugees into Somalia. The Somali regime claimed there were more than one million refugees from Ethiopia in Somalia in the mid-1980s. The situation was reversed in the late 1980s, when the revolt against the Siad Barre regime in northern Somalia led by the Somali National Movement (SNM) entered its final stage with rebel efforts to seize the towns. The contested towns were levelled by bombing from the air, and the bulk of the population of the northern region, several hundred thousands strong, fled across the border into the Ogaden region of Ethiopia, many of whose inhabitants were at that time refugees in Somalia.

The internecine war that broke out in southern Somalia after the fall of Siad Barre in 1991 pushed yet another wave of Somali refugees into Ethiopia, Djibouti and Kenya. The last two countries had not produced significant numbers of refugees until then, although they had played host to Ethiopian, Somali and Southern Sudanese refugees for several years. In the early 1990s, however, when civil war broke out in Djibouti between the Afar and the Issa Somali, Afar refugees from that state fled into Ethiopia.

'Economic refugees' are an acknowledged, though not officially recognized, category of involuntary migrants. Students of the subject debate the degree of actor volition involved, and the comparative weight of 'push' and 'pull' factors in the decision to migrate (see e.g. Mekuria Bulcha, 1988). At one end of the scale are those who are able to calculate the opportunity costs of such a move and decide accordingly. Such people usually have an urban background and possess some education and skills, as well as the sophistication to plan their move. Many such people left the Horn in the 1970s and 1980s, in particular from Eritrea, Ethiopia and Sudan. Their destination was seldom one of the neighbouring countries in the region, for these have little to offer in terms of economic opportunities. Ethiopians and Eritreans flocked to the United States, Canada and Europe. In Sudan, this type of refugee normally hails from the North and is to be found in the oil-rich Arab countries of the Gulf. Not a few Somali also went to the Gulf. In the early 1990s, the violent political storm that hit Somalia dispersed the Somali intelligentsia to the four corners of the world.

At the other end of the scale are those who have little choice in the matter, because they cannot survive in their homeland. People threatened by famine belong to this group, and they comprise by far the largest refugee category in the Horn. Famine is not indiscriminate. Amartya Sen (1981) rightly notes the characteristic feature of starvation is not that there is not enough food to eat, but that some people do not have enough to eat. Famine in the Horn is a rural phenomenon. It ravages peasants and pastoralists alike; if there are starving people to be found in the towns, they are mainly refugees from the countryside.

Famine is no stranger to the Horn. Traditionally its inhabitants had ways of coping with its visitations. Indigenous strategies of coping with the threat of famine are a favourite research theme for post-famine studies, which show that most such strategies are based on movement and temporary relocation to other areas. Such movements are facilitated by pre-existing relationships between guest and host communities. In recent decades, however, political conflict has inhibited movement and disturbed traditional relationships between communities in the Horn, leaving people defenceless in the great famines of the 1970s and 1980s and the smaller visitations in-between. During this period, the main coping strategy was also based on movement. In the great Ethiopian famine of the early 1970s, the starving people of Wolo and Tigray Provinces moved into the towns that line the main arterial road leading south. This is where they were discovered by television, and famine became a world 'media event'. Likewise, in the famine of the mid-1980s in Sudan, the starving people of Darfur moved east to the vicinity of Omdurman. In Ethiopia in the mid-1980s, the 'flight dynamics' were different. Cut off from the rest of Ethiopia by the war, starving people from Tigray and Eritrea were forced to trek over great distances to refugee camps in eastern Sudan. Rural refugees are likely to stop at the first place after they cross the border where they find relief, particularly if the area is inhabited by kindred peoples. Only in one instance did rural refugees find welcome farther abroad. These were the Falasha, the so-called 'Ethiopian Jews', who were taken to Israel.

Environmental refugees are the latest category of involuntary migrants, and the one likely to prove the most durable. These are people who are forced to abandon their homeland due to ecological degradation, with little hope of returning. People from famine-prone regions in the Horn can be properly considered environmental refugees, because these regions are believed to have reached a stage of advanced degradation and are no longer capable of supporting their population. These regions include parts of northern and eastern Ethiopia, most of Eritrea, western Sudan, northern Somalia and northern Kenya. Environmental refugees require permanent settlement in their place of refuge: this is one of the intractable problems confronting the United Nations High Commissioner for Refugees in the Horn (Gaim Kibreab, 1987). In some rare cases, refugees may be able to settle if land is available and the host community is accommodating. One such case involved the Beni Amer pastoralists from western Eritrea, who settled among their Beja kinsmen in Kassala Province of eastern Sudan.

In most cases environmental refugees are likely to form part of the drifting mass of 'internally displaced people', the last category of involuntary migrants to be found in the Horn. Their numbers run into millions, although they have never been counted. The reasons for their displacement can be political, economic, or environmental; and their plight is extreme because they seldom receive assistance. The worst instance of internal displacement is the exodus of Southern Sudanese, an estimated three million of whom were to be found in Northern Sudan in the early 1990s (Burr, 1993). Often, internally displaced people flock to urban centres in the hope of finding some means of livelihood. Addis Ababa doubled its population between 1970 and 1984, and quite likely doubled it again in the decade that followed. The Khartoum–Omdurman urban complex was surrounded by an expanse of slums inhabited by refugees from Southern Sudan and impoverished areas of the North. Its population quadrupled in the first 25 years of independence, and had reached 2.9 million in 1993 (Sudan, 1995, p. xiii).

The influx of a sizeable wave of refugees in any area is seldom without tension, and sometimes provokes conflict with the host community and government. The worst instance, a little-known one, concerns Somali refugees in Kenya's Northeastern Province. This area of Kenya has been under emergency rule since the Somali secessionist struggle of the mid-1960s. Its Somali pastoralist inhabitants are still viewed as potential subversives by the Kenyan authorities, and Somali refugees from across the border are unwelcome and often mistreated. With the influx of large numbers of such refugees in the early 1990s, there came reports of serious abuses against them by the Kenyan security forces (African Rights, 1993a).

Djibouti was nearly overwhelmed by refugees from Ethiopia and Somalia, whose combined numbers at one point equalled one-quarter of Djibouti's normal population. The influx commenced with the arrival of about 40,000 people from the Ogaden in 1977, in the wake of the war that had erupted there. They were followed by people fleeing political repression in Ethiopia. The Djibouti authorities initially refused to grant them formal asylum, and later sought to draw a line between 'genuine' and 'illegal' refugees. The former were assisted through the Office National d'Assistance aux Refugiés, while the latter were subject to deportation. When full-scale civil war broke out in Somalia in the late 1980s, a new wave of refugees flowed into Djibouti, creating a crisis of rising prices and shortages of basic foodstuffs. Moreover, these refugees were a potentially unsettling factor in the domestic politics of the host state, poised precariously, as they are, on a delicate balance between two ethnic groups – Somali and Afar. The authorities in Djibouti were anxious to get rid of them, and considerable pressure was exerted on many refugees to return to their homeland or move on to a third country.

Sudan set up its own Office of the Commissioner for Refugees in 1968, and refugees in that country were treated with tolerance despite their large numbers. However, the same was not true of internally displaced Sudanese who congregated around the urban conurbation of Khartoum and Omdurman (El Tigani, 1995). From 1979 onwards, successive Sudanese regimes resorted to the infamous practice known as *kasha*: the forcible mass expulsion of displaced persons from the vicinity of the capital region. The famine of the mid-1980s forced large numbers of starving people from Darfur in the west to trek vast distances to reach Omdurman in the hope of finding assistance. The Nimeiry regime reacted by sending them forcibly back to where they had come from. Nevertheless, the refugee slums in the capital region continued to grow as the war in the South intensified, forcing hundreds of thousands of Southerners to leave their homes, and the authorities began to worry about the security implications of a large congregation of Southerners near the capital. By the early 1990s, an estimated 1.5 million people lived in appalling conditions in the area. The military regime that came to power in Khartoum in 1989 proceeded to implement the expulsion programme with zeal, destroying slum housing and removing people by force. In March 1992, it was announced that one million people had been expelled (*Africa Watch*, 1992).

A rather different political problem emerged in Somalia with the massive influx of refugees from the Ethiopian Ogaden after 1977, and smaller numbers during the famine of the mid-1980s. Since most of them were ethnically Somali, according to the constitution of the Somali Republic they were entitled to citizenship, and the government of Somalia did not consider them aliens. This was not without political implications. The bulk of the refugees belonged to the Ogaden clan, one of the pillars of support for the Siad Barre regime. The regime was then engaged in a mortal struggle with the Ishaq clan family in northern Somalia, and the Ogaden had been feuding with the Ishaq for decades. In the event, many Ogaden men were enlisted in the exceedingly violent campaign of repression waged against the Ishaq in northern Somalia in the late 1980s. Ironically, this campaign forced hundreds of thousands of Ishaq to seek refuge in the Ogaden homeland inside the Ethiopian border.

Refugees can also create economic contradictions and conflict with the host community due to their impact on local resources. The presence of large refugee camps has an immediate and dramatic impact on the local environment. Trees and shrubs in the vicinity disappear, water resources diminish and become polluted, the ground is eroded, and the area turns into a dust bowl.

Prices may rise due to increased demand for basic goods, if the refugees are able to shop in the market: a development unwelcome to the local community. Or, it could mean falling prices, if goods are distributed gratis to the refugees, who then channel a portion of them to the market. In such a situation, local production of foodstuffs may be discouraged. In Somalia for example, refugee ration cards became the principal means of exchange. The impact on the labour market can be serious also, if the refugees compete with local workers, and this may work to the disadvantage of both groups. For example, refugees in Djibouti were not allowed to work without special permission. Those who did obtain it were paid at half the rate commanded by Djibouti nationals.

The influx of large numbers of refugees may also have an impact on the distribution and maintenance of the most basic of all resources, land. That in turn can cause conflict. During 1984–88, about 600,000 people from northern Ethiopia were resettled in other regions, mostly in the southern part of the country. Although the overall impact in the resettlement region was not great, the increase in population density in some of the actual areas of settlement was significant (Pankhurst, 1992). Increased population pressure on land, water and forest resources disrupts traditional agricultural practices, such as fallowing patterns, with unpredictable ecological consequences. When pastoralists are displaced they bring along as many animals as they can save, and the animal population density in the host area multiplies suddenly, with overgrazing and degradation as the result. In 1981, the Somali government claimed refugees from Ethiopia had brought with them about three million undernourished cattle, posing a threat to the national herd and to an environment already weakened by prolonged drought (UNDP, 1981).

Since the size of the national herd of cattle in Somalia was estimated at about four million in the 1980s, the figure of a three million influx seems dubious. As a rule, statistics relating to refugees and displaced persons in the Horn are estimates of variable quality, and must be taken only as indicators of the order of magnitude involved. Political considerations are often involved, and economic calculations may also play a role. For example, the size of the refugee community in Somalia in the 1980s became a game of numbers played by the government of that state and the international aid agencies. The former put forth highly inflated figures, and deflected all attempts by the latter to carry out some form of enumeration. The Somali government was seeking to maximize the amount of assistance coming from abroad which, in fact, became a major source of revenue for the Somali state during that decade. Not only governments, but also armed opposition movements are prone to exaggerate the needy population within the territory they control, in order to increase the amount of aid allocated to their area. The political element is not always fairly accounted for in analyses of famine and population displacement. As one observer notes, what tends to be missing 'is any sense of the function that violence may serve, or of how various powerful groups (perhaps including the national government) may have an interest in actively promoting violence and famine for purposes of their own' (Keen, 1992, p. 28). A familiar example is provided by embattled regimes blocking access to food for people in rebel-held areas in order to starve them into submission, while rebel movements seek to starve those in government-held areas into surrendering.

1.4 Summing Up

In the Horn, population movement and conflict are obviously related, but the nature of the relationship is not obvious. It might seem that conflict is the cause of movement, but the reverse can also be true. In the former instance, the effect is immediate and dramatic and the relationship easy to perceive. In the latter, the process is cumulative and hard to perceive. Other factors may intervene to make the connection even more obscure. The region's resources are meagre and unevenly distributed, and its history is a record of frequent population movement: a mode of adaptation to, rather than of, the environment. More often than not, movement is caused by a shift in the balance between people, animals and natural resources, and this is dramatically manifested when the balance is suddenly and drastically upset; for example, by drought. Demographic, environmental and economic trends in the Horn during the past decades have undermined the sustainability of many local production systems, forcing people to seek alternatives or supplements abroad: through emigration, seasonal migrant labour, trade, or relocation to an urban slum. Such shifts are gradual and hardly noticed, until they give rise to conflict.

All these trends can be expected to accelerate in the future, and population movement in the Horn is unlikely to subside. According to one study, the recent history of the Horn suggests that population movements are an integral part rather than an incidental element of the regional political economy' (Crisp & Cater, 1990, p. 20). In a region of scarce resources, population movement is likely to provoke conflict because it puts additional pressure on an already fragile economic base. Itself the product of scarcity and instability, population movement will often lead to more conflict and instability. There are numerous instances of this in the recent history of the Horn, and many will be described in the pages that follow. One example is given below because it illustrates clearly the relationship examined here, despite the intervention of various other factors.

In the second half of the 1980s, the Darfur region of Sudan was engulfed in a violent many-sided conflict in which not only Sudanese but Chadians and Libyans as well became involved. The Sudanese government inveighed against 'bandits' and 'foreigners' and such elements were present, but the essence of the dispute was land, and it involved mainly local groups (Sharif Harir, 1993). The homeland of the sedentary Fur people, with the Jebel Marra massif as its core, is relatively well endowed, and it attracted numerous environmentally and politically displaced groups in the 1970s and 1980s. The latter were Arab-speaking groups from neighbouring Chad, a country mired in unending civil war. The former were Zaghawa nomads from northern

Darfur, an arid region completely dry two out of every four years. The Zaghawa regularly crossed Fur territory in dry years to reach pastures to the south, negotiating their rights of passage and land use with the Fur traditional authorities.

Nomads from Chad, kinsmen of the Zaghawa, also entered Zaghawa territory with their animals. This additional pressure did not allow the rangelands time to regenerate. Northern Darfur was hit by the Sahelian drought in the 1970s, and by the middle of the decade a sizeable population movement to the south into Fur territory was taking place. The movement intensified when drought worsened in the early 1980s, and the nomads suffered heavy losses of livestock. Deprived of their traditional means of livelihood, they sought land to cultivate. In this they were joined by other groups fleeing the drought, and competition over land in the Fur region became intense, leading to conflict. This took an ethnic form, with the Arab groups from both sides of the border joining forces against the Fur and other non-Arab groups. Political factors also intervened. The resumption of the conflict in the South prevented movement southwards. The Nimeiry regime in Khartoum had earlier abolished the native administration system dating from colonial times, and the traditional system of settling inter-communal disputes no longer functioned. In 1981, the regime implemented a regional self-administration scheme, and the administration of Darfur was headed by Fur people. The army and police under this administration took harsh measures against the Zaghawa, provoking them into rebellion. Tribal militias were formed by all groups, and all-out war ensued. It was not until the end of the decade that the central government was able to re-establish control.

2

Regional Resources

2.1 Introduction

People, it has been said, are the 'ultimate resource' (Simon, 1981). Indeed, the history of the Horn shows its people have been resourceful and adaptable. Within the limits of their environment and the technology available, they created sophisticated material cultures and produced regular surpluses to support complex societies and large states. The Abyssinians worked the land intensively with oxen and the iron-tipped plough, and the Northern Sudanese used basin and lift irrigation in the Nile valley. Indigenous food-producing plants were developed, like *teff* and *ensete* on the Ethiopian plateau, *dukhn* in Northern Sudan; and native products like coffee, gum arabic, myrrh and frankincense reached the international markets. The vitality of the regional economy was manifested in the far-flung activities of its famous trader groups. The *Jallaba* of Sudan ranged from Central Africa to the Red Sea and from Egypt to the swamps of Southern Sudan, while the Abyssinian *Jabarti* traversed the plateau and traded on the Red Sea coast. The trading towns on the coast of the Red Sea and the Indian Ocean were integrated into an international commercial network linking Africa to Europe and Asia.

Agriculture has always been the mainstay of the region's economy, and it combines cultivation and animal raising. The people of the Horn till the land where it is possible to do so, and raise livestock where it is not, usually managing to do a bit of both. Given the elementary level of the technology used, agriculture in this part of the world is subject to conditions set by an untamed and capricious nature. The natural resources that sustain agricultural production are the most precious assets for the people of the region. This chapter will examine the natural endowment of the Horn, with particular attention to basic resources that have come under stress due to excessive demand and have become actual or potential sources of conflict. We begin with a physical profile of the region.

2.2 Relief

The Horn contains sharply contrasting physical regions. The most pronounced contrast is between the massive Ethiopian–Eritrean plateau at the centre and the lowlands that surround it. The shape of the plateau is a triangle inclining towards the west and south. It has a mean elevation of 2,000 to 3,000 metres and peaks that soar to over 4,000. This is a highly fragmented massif with towering peaks and sunken valleys, and elevations ranging from 1,700 to 4,600 metres. Traversed by deep and wide canyons and cut by many rivers that pose formidable obstacles to transport and communication, the plateau is split into two uneven sections (eastern and western) by the Great Rift. The Rift itself is lined with a string of lakes. Given its highly dissected nature, the plateau has few flat areas. Unusually, human settlement is found on hilltops rather than in low-lying, flat areas, and cultivation is often done on slopes.

The plateau towers over the adjacent plains that stretch towards the sea in the east and south and to the Nile valley in the west. In the northeast, now part of Eritrea, the drop to sea level is precipitous, and the Red Sea plain is a mere strip that widens progressively southwards. In parts of the Danakil Depression, the plain sinks below sea level. In the southeast and south the land descends gradually towards the Gulf of Aden and the Indian Ocean, reaching its lowest level in the Somali Republic. In the southwest it slopes downwards till it reaches northern Kenya, then begins its ascent towards the East African highlands.

The plateau overlooks the immense Sudanese Nile basin, crossed by the mighty river that flows within Sudan for more than 4,000 kilometres. The Nile valley is the broad bottom of a basin that slopes gently northwards, surrounded on three sides by highlands. The average elevation of the valley is 300–400 metres. The gentleness of the slope makes for a low velocity of water flow in the river, encouraging vegetation growth as well as high rates of evaporation. A vast area of nearly 5,000 square kilometres contains the plant-choked swamp known as the Sudd, which effectively divides Sudan into two quite distinct and unequal parts, North and South. Southern Sudan comprises about one-third of the country's total area.

The Golis (Ogo) mountain range, with peaks rising to 2,500 metres, is the eastern extension of the plateau in northern Somalia. Three-quarters of the vast Haud plain that descends from there southwards fall into Ethiopia and the rest in Somalia. Lowlands stretching southwards from Ethiopia and westwards from Somalia make up the bulk of Kenya's territory, providing geographic, ethnographic, environmental and economic links between the three states. The northern Kenyan lowland includes the Chalbi Desert and volcanic peaks that rise to 2,000 metres. The Kenyan highlands are also bisected by the Great Rift, which runs north to south through their centre. These highlands have several volcanic peaks over 4,000 metres, including majestic Mt Kilimanjaro (5,895 m), the highest in Africa. The Rift has several lakes, the largest being Lake Victoria, which lies at an altitude of 1,134 metres.

2.3 Soils, Vegetation, Land Use

The geological composition of the Horn consists of two structural levels. The lower is composed of Precambrian and Lower Palaeozoic crystalline and metamorphic rocks, and the upper predominantly of Mesozoic sediments and Cenozoic sedimentary and volcanic rocks. The types of soil found in the region are not particularly well suited for agriculture. A survey of the Ethiopian–Eritrean plateau and surrounding lowlands indicated that only 25% of its soils are good for cultivation, more than 30% are useless, and the rest have limited agricultural potential (*National Atlas of Ethiopia*, 1988, p. 8). Old and low alkaline soils predominate in the higher and cooler central highlands, old and highly acidic soils in the lower and humid southwest, and relatively new, low acidic and salty soils in warm areas of moderate rainfall in the northeast and southeast. Great piles of volcanic lava form the mountains of the region.

The soils of the Sudanese Nile basin are sandy in the north, stabilized sand dune (*qoz*) and clayey in the central part, and lateritic in the south. The sands of the north are infertile, while the environment in the south severely limits the usefulness of its soil. Only the *qoz* and clay plains of the central region have good potential for agriculture. Alluvial soils are the most valuable. In Sudan these are found in the inland deltas of the Barka and Gash Rivers, along the White Nile north of Malakal, the Blue Nile north of Roseires, the main Nile north of Khartoum, and in the valleys of the rivers radiating from the Jebel Mara mountains in Darfur. Called *seluqa,* such land is exposed when river flow is low.

On the Shebeli River, alluvial flats one to five kilometres wide begin at Ime in Ethiopia and widen southwards, making the river valley in southern Somalia fertile. Alluvial flats are found also along the Juba River. The Haud plateau is limestone, and Somalia's coastal zone has unstable, sandy soils with limestone deposits. In Kenya, parts of the highlands have rich volcanic soil with good humic content. The best soils are the red clay loams of the Kikuyu region. The Lake Victoria basin also has fertile sandy, loamy soils. Elsewhere in Kenya the soils are leached, alkaline and poor. The Rift Valley has thin soils evolving from alkali lavas and ashes. The lowlands of northern Kenya have shallow soils that are often stony or alkaline.

Vegetation cover in the Horn is determined primarily by the climate, especially precipitation, and secondarily by human and animal activity. In the lowlands, precipitation is low and erratic. Consequently, the vegetation cover is generally sparse, and includes wooded savannah in the better watered areas, bushes and shrubs in the drier areas, and only seasonal grasses in the arid zone. Unimproved grazing by mobile pastoralists is the dominant form of land use in the lowlands, except where rivers permit intensive cultivation through irrigation. These two forms of land use do not coexist easily, because irrigated cultivation deprives pastoralists of their most valuable pastureland. In the Horn this contradiction has become a source of conflict. Intensive cultivation is also practised in the savannah lands of central Sudan, where modest precipitation permits the exploitation of the *qoz* and clay plains. Here the impact of man has become obvious, with the rapid disappearance of wooded cover.

Forest cover is scarce in the Horn. Even on the Ethiopian plateau, where it must have been plentiful once, forests now cover less than 4% of the land, and that only in the southern part. A larger part of the plateau is woodland (deciduous, juniperous, acacia), and the rest is savannah. Southern Sudan with its unmodified tropical climate and seasonal flooding has about 350,000 square kilometres of forest. Scattered acacia bush and seasonal grasses form the vegetation cover of the northern half of the Nile basin in Sudan. Here no cultivation is possible due to inadequate precipitation, save on narrow irrigated strips along the river banks. Savannah cover with gum arabic and acacia trees is typical of the central region of Sudan, where intensive rain-fed and irrigated cultivation is changing the face of the land. In northern Somalia, remnants of juniper and evergreen woodlands can be found on the higher reaches of the Ogo mountain range. Southern Somalia is savannah land, but remnants of tall trees can be found in the alluvial flats along the Juba River. There is limited mangrove growth in Somalia along the extreme south of the Indian Ocean coast, near the border with Kenya.

In Kenya, forests cover 3.4% of the total land area. Since they are situated in highland regions of great agricultural potential, they are gradually being cleared for cultivation. On the south and eastern slopes of Mt Kenya, for instance, between altitudes of 1,400 and 2,100 metres, the forest has been cleared for cultivation. Three-quarters of the total area of Kenya is covered by rangeland types of vegetation, including woodland, bushland, shrubland and grassland. The arid zone, a great arc from the southeast to the north and northwest, is rock-strewn shrubland, interspersed with hilly outcrops and mountains. The grassland savannah of south central Kenya includes the game parks of Amboseli, Masai Mara and Nairobi. Kenya's coastal strip, 10–40 miles inland, is dotted with dunes and coconut groves. Its estuaries host extensive mangrove forests, in all some 65,000 hectares.

2.4 Climate

In the words of one scientist (Griffiths, 1972, p. 134), the Horn 'presents a strange climatological phenomenon' because, although it lies in the equatorial tropical zone, it suffers from serious water deficiency: a climatic anomaly not encountered in similar latitudinal and geographic locations. The Horn shares this deficiency with the wider East African region which, according to another scientist, has one of the 'problem climates' of the world (Trewartha,1961). Part of the explanation for this anomaly lies in the fact that 'the prevailing winds during most months of the year have a northeasterly or southwesterly direction, thus making moist air masses over the land an exception rather than the rule' (Griffiths, 1972, p. 134). The greater part of the Horn lowland receives a mean annual rainfall of less than 500 mm and has a moisture index of minus 40 to minus 50.

By contrast, the plateau has a typical highland tropical climate, with heavy precipitation in summer and autumn, and some areas receiving rain in winter as well. Precipitation ranges from 1,000 mm to 2,000 mm per annum. The southwestern section receives rain during three-fourths of the year and averages 1,500 mm to over 2,000 mm. The precipitation rate decreases towards the northeast and east. Annual rainfall in Addis Ababa during 1900–1959 ranged between 992 mm and 1,905 mm (Griffiths, 1972, p. 375). In Tigray and Eritrea in the north it drops to an annual mean of 500 mm. To some degree, precipitation increases with elevation within the plateau, but its pattern is erratic throughout the Horn. The impact of this unreliability is felt mainly in the areas of low rainfall. In areas of abundant rain, a deviation from the mean of even up to 30% does little damage, while the drier lowlands can be devastated by a deviation of even 5%.

On the plateau, temperature correlates with elevation, the approximate change being 0.7° Celsius per 100 metres. Frost is not uncommon in the higher reaches, although even the highest peak (Ras Dejen, 4,620 metres) has no permanent snow cap. The traditional Ethiopian land classification has five categories: *woorch* are the coldest heights above 3,500 metres; *dega* are the cool highlands above 2,500 metres, the historic home of the Abyssinians; *woyna dega* are the warm lands between 1,500 and 2,500 metres; *kolla* is the hot zone below 1,500 metres; and *haroor* is the scorched zone below 500 metres. The lowlands of the Horn have very high temperatures: it was in the Dallol area of the Danakil Depression that the highest ever mean annual temperature on earth – 34°C – has been recorded.

The Nile basin in Sudan is wholly tropical. The rate of precipitation over its length varies greatly, ranging from less than 100 mm in the arid far north to 1,800 mm in the south. Since there are no mountains to interfere with air streams, gradation is smooth and no obvious dividing line occurs. The Sudanese distinguish four climatic zones which correlate with vegetation regions. The far north is the desert region with less than 100 mm rainfall. South of this is an arid, semi-desert region with 100–300 mm. These two regions cover about half of Sudan's entire territory. No cultivation, except along river banks, is practised here. To the south lies the savannah woodland between latitudes 9° to 15° North, covering the width of the country. Rainfall of 300–800 mm makes cultivation possible here, in what is considered the most important agricultural region of Sudan. Southern Sudan is inundated part of the year, with a rainfall rate ranging from 700–1,800 mm and heavy seasonal flooding of its many rivers.

The climate of Somalia is governed by the monsoon winds and oceanic current systems. There are two rainy seasons, spring and autumn. Precipitation is generally low, irregular and unevenly distributed, and increases from north to south. It averages 500 mm to 600 mm in the southwestern part, diminishing to less than 100 in the northeast. With as much as 500 mm of rainfall, the heights of the Ogo range are far better watered than the Guban, which receives less than 100 mm.

Kenya also has two rainy seasons, spring and autumn. There are wide differences between the high and low lands. About 72% of Kenya's total area receives less than 500 mm, and only 15% receives over 750 mm. The largest part of the northern lowland receives less than 250 mm. There is moderate variation in temperature between low and high land, the general rate of decrease with altitude being approximately 5.3° Celsius per 1,000 metres. Kenya sits astride the Equator, but only its narrow coastal strip is tropical; 18% of the land is highland steppe with a temperate climate.

A very large part of the Horn is classified as arid. 'Aridity is an expression of water deficiency: and water deficiency is induced not only by lack of precipitation but also conditions of soil moisture and permeability, evaporation, transpiration by plants, and the intensity and duration of sunlight, heat, humidity, and wind' (Hodge & Duisberg, 1963, p. 21). The most frequently used measure of aridity is the moisture index, which is derived from a formula that compares precipitation with evaporation and plant transpiration. The moisture index is zero when precipitation = evaporation + plant transpiration. It is minus when there is less precipitation, and plus when there is more. Arid regions are characterized by rainfall that is insufficient to replenish loss of moisture. Generally the moisture index for arid land is below –40, while semi-arid land has an index of –20 to –40. A simpler indicator is precipitation alone: less than 500 mm makes for aridity, 500 mm to 750 mm for semi-aridity.

The moisture index expresses the relationship between water supply and water need. Here it should be noted that soil and climatic conditions in the Horn make for a high water requirement. While in Europe a mean annual precipitation rate of 750 mm is sufficient for most crops, it is generally insufficient for the Horn of Africa. According to some, the dividing line where continuous cultivation ends and rangeland begins in East Africa is 762 mm of precipitation (Peberdy, 1963, p. 158). This excludes so-called semi-arid terrain (500–750 mm), parts of which can be cultivated, as is the case with the central region of Sudan. Generally, aridity is said to prevail below 500 mm. According to this criterion, the entire coastal zone of the Horn is arid, as is northern Somalia, save for the northwestern corner, and southern Somalia save for the inter-riverine valley. Three-quarters of Eritrea and Kenya fall into this category, more than half of Sudan, all of Djibouti and the eastern and southeastern lowlands of Ethiopia (see Table 2.1).

Extreme aridity (0–100 mm) turns land into desert. There is no precise definition of this phenomenon, though several criteria can be used to devise one, including rainfall, vegetation, soils etc. Generally, it connotes sandy soils, the absence of vascular plants and the existence of specially adapted animals. The northern quarter of Sudan is an extension of the Libyan and Nubian deserts. Most of the Danakil region in Eritrea and Ethiopia is classified as desert, so is part of southeastern Ethiopia, and part of Kenya's northern lowland. While desert in origin is a natural phenomenon, 'desertification' is an extension of it through human intervention. Such intervention occurs when cultivation is practised in the arid and semi-arid zones, resulting sequentially in the reduction of vegetative cover, exposure of the soil to accelerated water and wind erosion, reduction of soil organic matter and nutrient content, deterioration of soil structure and hydrological properties, loss of biological productivity, with crushing and compaction leading to further loss of soil fertility.

Table 2.1 *Land*

Country	Total area (sq. km)	Arid land (% of total)	Cropland (% of total)
Djibouti	23,000	100	0
Eritrea	125,000	75	–
Ethiopia	1,098,000	52	13
Kenya	582,646	72	4
Somalia	637,657	75	2
Sudan	2,505,813	66	5

Sources: Country statistical data; World Bank, 1994, 1995. *World Development Report*. Washington, DC.

The Horn features prominently on all 'desertification' maps, although there is no agreement on what this process entails, or indeed, whether it is actually occurring. 'Desertification' evolved into a fuzzy concept in the wake of the great Sahelian drought of 1968–73. It was credited alternatively to advancing desert margins (desertization), to prolonged drought (desiccation), and to human and animal activity (environmental degradation). The four alleged principal causes of land degradation are over-grazing on rangelands, over-cultivation on croplands, deforestation and soil salinization: all familiar phenomena in the Horn. The United Nations Conference on Desertification (UNCOD, 1977) in Nairobi cobbled together the following definition:

> Desertification is the diminution or destruction of the biological potential of the land, and can lead to ultimately desert-like conditions. It is an aspect of the widespread deterioration of ecosystems and has diminished or destroyed the biological potential, i.e. the plant and animal production, for multiple-use purposes at a time when increased productivity is needed to support growing populations in quest of development.

In the follow-up to the 1992 United Nations Conference on Environment and Development, a committee on desertification was established, ponderously named the Inter-Governmental Negotiating Committee for the Elaboration of an International Convention to Combat Desertification in Those Countries Experiencing Serious Drought and/or Desertification Particularly in Africa. The same year, the United Nations Environment Programme published the *World Atlas of Desertification* (UNEP, 1992a), which carried the dire warning that 'one sixth of the world's population is threatened by the effects of desertification'. Only two years later the credibility of these claims was challenged by none other than the two co-authors of the *Atlas* itself, who cast doubt on the data upon which these claims were based, and denounced the desertification notion as a myth (Thomas & Middleton, 1994).

Northern Sudan is said to be menaced by advancing deserts. An aerial survey of the desert margins in Sudan was completed in 1975, and the photographs were compared with maps prepared in 1958. One expert concluded from this evidence that the desert had shifted south 90–100 kilometres in 17 years, and was advancing at the rate of 5–6 kilometres per year (DECARP, 1976, p. 11). This finding was used to solicit international aid to 'stop the desert'. The methodology used was challenged by others (Hellden, 1984), who believe the desert advances and retreats following variations in rainfall and grazing pressures. Northern Ethiopia is said to be losing soil at fantastic rates. An official report cited a figure of 32 tons per second, and this became something of a benchmark for the Ethiopian predicament (Tadesse Kidane Mariam, 1985, p. 4).

'Desertification' is complexly related to climatic change, both as cause and effect. Whether or not such a change has been taking place in the Horn is the subject of a longstanding but inconclusive debate. There are reports of rainfall decline this century in the tropical zone worldwide (Kraus, 1955), in the Sahel (Wistanley, 1973), in the Ethiopian drainage basin (Butzer, 1961) and in Aden (Hemming, 1966). There are also reports of volume reduction in the Nile flow (Hemming, 1966), and falling levels in Lake Turkana and Lake Stephanie (Butzer, 1971). These are contradicted by reports based on data collected at Serpent in Djibouti that show no trend of rainfall reduction since 1901 (Griffiths, 1972, p. 135). According to one calculation, the 1968–87 period in the western and central Sahel was significantly drier than the 1931–60 period, but this does not apply to the Horn (IUCN, 1989).

One school of thought maintains that persistent drought may have local causes, such as loss of vegetation cover, which in turn changes the albedo (reflectivity) of the ground surface and soil moisture. Loss of vegetation increases the albedo and this, in turn, changes the heat balance of the surface-atmosphere system. Large-scale changes in heat sources result in reduced uplift in the higher albedo region. Less uplift means less rain and, consequently, less vegetation. Loss of vegetation results in reduced moisture retention in the soil. Less soil moisture means less evaporation which, in turn, results in reduced upward motion and convective activity and less rain. These are assumed to be bio-geophysical mechanisms which, once started, feed back on themselves to perpetuate drought conditions (see Rasool, 1984).

Drought is a familiar visitor to the Horn. Its visits are well remembered for they are usually accompanied by famine. A rough collation of recorded incidents this century suggests major drought occurs every ten years, with localized incidents occurring much more frequently. There is no evidence that a macro-climatic, i.e. permanent, change is taking place in the Horn, though it could well be that a micro-climatic, i.e. reversible, phenomenon is in progress. Those who absolve climate as a cause of 'desertification' maintain that human and animal activities – over-grazing, over-cultivation, deforestation, salinization – have increased the vulnerability of land to the vagaries of the climate.

2.5 Fresh Water

In a predominantly arid region, fresh water is a precious resource because it can bring land into cultivation through irrigation. Because it is precious, fresh water is the object of competing and conflicting demands among various parties: subsistence cultivators versus commercial farmers, farmers versus pastoralists, the energy industry versus the tourist industry, and environmentalists versus all the rest. Competition and conflict occur at various levels within and between states. Since fresh water is a national resource, its disposition among the various claimants within state boundaries is determined by the state itself; that is, by those who control state power. The reaction of those who are adversely affected in this process may lead to conflict, though seldom in direct fashion. For example, resentment over the loss of valuable pasturelands to irrigated farming certainly is a major contributing factor to the widespread political alienation of pastoralists in the Horn, and is indirectly linked to their involvement in endless conflict; though the matter is seldom presented in such simple terms. Increasing water scarcity in the future is forecast for the region. 'Water stress' has been predicted for all the countries of the Horn by Falkenmark (1991, p. 93), who believes water scarcity will probably affect demographic trends in a way that will confound current predictions of population growth patterns. Sudan, Kenya and Somalia are among countries predicted to have water shortage in the year 2000 (FAO, 1993c, p. 238).

The potential for conflict between states in the Horn over fresh water is very real, because the states which are the source of the water have as yet made little use of it, and any use they make of it in the future will have a serious effect on downstream states which are critically dependent on irrigation. Utilization of fresh water already figures importantly in the development plans of upstream states, and these have aroused concern in the states downstream. So far, attempts to promote regional cooperation in this highly sensitive area have proved fruitless. Unless there can be progress in the near future toward cooperation in the sharing of freshwater resources, interstate conflict in this region is a real possibility.

The main sources of fresh water for the region are the Ethiopian–Eritrean plateau and the East African highlands. The plateau is a water divide between the Mediterranean Sea and the Indian Ocean and the water reservoir of the Horn; a fact of great geopolitical significance and one fraught with the potential for serious conflict. The plateau is drained by eleven major rivers, which take nearly all but an estimated 3% of the run-off water to the surrounding lowlands. Given the westward tilt of the plateau, the larger western section drains into the Nile basin and the water is carried by the Nile to the Mediterranean. The smaller eastern section of the plateau drains southward, but only a small portion of the water ever reaches the Indian Ocean. There is only minor drainage to the east, and little water reaches the Red Sea.

The prime beneficiary of the drainage pattern in the Horn is Sudan, and the bulk of the water comes from the Ethiopian–Eritrean plateau. The principal contributor is the Blue Nile which drains the central plateau. It starts at Lake Tana in northern Ethiopia and meets the White Nile at Khartoum, accounting for about 60% of the total discharge of the main Nile. The Takaze (Setit) River drains the northern section of the plateau and joins the Atbara River in Sudan. The Gash (Mareb) and Barka Rivers rise in the Eritrean section of the plateau and flow into eastern Sudan; the Gash River seasonally fans out north of Kassala, fertilizing an area of 100–150 square kilometres, while the Barka forms a delta at Tokar on the Red Sea coast.

The second source of the Nile is the Equatorial Lakes, where the Upper Nile begins its long journey to the Mediterranean. In Southern Sudan it is joined by the Bahr el Ghazal River, which rises in the Ironstone Plateau on the Sudan–Congo (Zaire) border, and the Sobat River, itself a confluence of the Baro River that rises in southwestern Ethiopia and the Pibor River which is fed by streams from the same region. All told, the rivers from the plateau provide about 87% of the main Nile's discharge. The world's longest river (6,650 km, if measured from its remotest headwaters in the Kagera River in Rwanda), the Nile runs more than three-fifths of its placid course within Sudan.

Somalia is the exclusive beneficiary of the drainage system from the eastern section of the Ethiopian plateau. The Shebeli River rises in the Harar range and has the largest (260,000 square kilometres) catchment area of all the rivers that drain the plateau. It travels 2,000 kilometres but fails to reach the Indian Ocean. Though usually classified as perennial, the Shebeli runs dry in its lower section part of the year. Three rivers that drain the eastern section of the plateau – Genale, Dawa, Weyb – join at the Ethiopia–Somalia border to form the Juba River, which crosses southern Somalia to reach the Indian Ocean near Kisimayo. These two rivers provide the only source of year-round flow of fresh water in Somalia and are the mainstay of its agriculture.

Only two rivers end within Ethiopian territory: the Awash and the Omo. The Awash, the only river that flows eastward, runs through the Rift Valley and ends up in Lake Abe on the Ethiopia–Djibouti border. The Omo flows southward through the Rift Valley, ending up in Lake Turkana on the Ethiopia–Kenya border. The Rift Valley floor is studded with a string of lakes with alkaline waters and rich animal life. Smaller lakes in volcanic craters are found throughout the plateau.

The major drainage channel from the East African highlands into the Horn is the White Nile as it leaves Lake Victoria, the world's second largest lake with a surface of 43,000 square kilometres and depths not exceeding 82 metres. The lake's drainage area is twice as large as itself and has high rates of precipitation, often exceeding 1,000 mm. The evaporation rate is also very high. It is estimated that only 15% of the water entering the lake leaves through the White Nile (Morgan, 1973, p. 260). Kenya contributes through six rivers to Lake Victoria. One major river and three lesser ones drain the Kenyan highlands on the east. The Tana, Kenya's largest river, rises near Mt Kenya, gets its water from that peak and the Aberdares Mountains, but loses most of it by seepage and evaporation as it descends onto the lowlands and flows south 800 kilometres to the Indian Ocean. The Ewaso Nyiro River has a similar origin, and its water dissipates in the Lorian swamp. The Athi and Sabaki Rivers leave the highlands and reach the Indian Ocean.

Large bodies of fresh water crossing arid regions inspire visions of 'greening the desert' through irrigation. In the past, irrigation by natural flooding, gravity and lift methods was used mainly along the Nile in Northern Sudan, while the first two methods were also used to a limited extent along the lower Shebeli in Somalia and the Baro in southwestern Ethiopia. Large-scale controlled irrigation was introduced during the colonial period in Sudan to meet the demand for cheap cotton from Britain's textile industry. The prototype was the immense Gezira scheme, launched in the 1920s on the Blue Nile. Using gravity-flow irrigation, it covered over 400,000 hectares by the time of Sudan's independence in 1956. Smaller schemes were introduced in the inland deltas of the Gash River in Kassala and the Barka River at Tokar. Mechanical pump irrigation was introduced along the banks of the Nile and some of its tributaries. Cotton became the dominant product of the colonial economy in Sudan, accounting for 62.4% of all export value in 1956. Irrigation was introduced during the same period by the Italians in the Shebeli–Juba valley in southern Somalia to produce bananas, cotton, sugar and other crops. Bananas became this colony's principal export.

As nearly everywhere in Africa, the post-colonial state in the Horn retained the colonial blueprint for the economy. The expansion of export-crop production was considered the shortest route to economic development, and irrigation the most efficient method for achieving it. In Sudan, the Gezira scheme soon doubled in size, and new schemes were founded in Rahad, Khasm el Girba, Kenana and elsewhere. In Somalia, several small irrigation projects were started in the Juba–Shebeli valley, and a large one at Bardhere on the Juba was planned to quadruple the area under irrigation. Irrigation was also introduced for cotton and sugar production in the Awash valley of Ethiopia, and along the Tana and Ewaso Nyiro Rivers in northern Kenya.

Irrigation has had a manifold impact in the region. It is the principal method for the production of many commercial crops (including cotton, sugar, fruit, etc.) which have become the mainstay of national economies and major sources of state revenue and foreign exchange. This is only the beginning of what is envisaged for the future. Irrigation is seen as the most effective means of ensuring food security. The following statement in Somalia's *Three Year Development Plan, 1979–1981* is typical: 'Continuing heavy investment in major irrigation work is inevitable since low and erratic rainfall is the greatest natural limiting factor to higher and more regular production' (Somalia, 1979, p. 88). Kenya's *Development Plan, 1989–1993* declares that 'irrigation development will make a major contribution towards the attainment of the objectives' for agricultural growth (Kenya, 1989, p. 130). Ethiopia, an official statement has asserted, 'cannot satisfy the food demand of her rapidly increasing population unless measures are taken to develop significant irrigation schemes in the near future' (Zewdie Abate, 1991, p. 8).

Plans for extending existing schemes and creating new ones are announced regularly in the Horn, where all countries claim to have extensive potential for irrigation. With some 1.6 million hectares under irrigation, Sudan is by far the heaviest user of freshwater resources in the region, but claims it is utilizing only half of its potential. Somalia, with 50,000 hectares under controlled irrigation, deems it possible to irrigate five times as much. With some 40,000 hectares under irrigation, Kenya finds it is using only 4% of its potential.

Hydroelectric power generation is another claimant of freshwater resources. Energy costs escalated rapidly due to oil-price rises in the past decades, making hydroelectric production a more attractive alternative. Environmental concern regarding fuel consumption also motivated donor agencies to favour hydroelectric development. All the states in the Horn except Eritrea generate electricity in this manner, but their needs far outstrip current production. The region's potential capacity for power generation is great. Current production by all countries in the Nile basin, it is estimated, represents only about 18% of potential capacity.

Ethiopia, where most of the water comes from, has made little use of it so far. Only the Awash River has been exploited for power generation and irrigation. Approximately 100,000 hectares came under irrigation for sugar and cotton production, while the potential irrigable land is estimated at three million hectares. The military regime that ruled Ethiopia from 1974 to 1991 made grandiose plans for power generation and irrigated cultivation. At the United Nations Water Conference (1977, Mar Plata, Argentina) it announced plans to irrigate large areas in the river basins of the Blue Nile and the Baro. At the United Nations Conference on the Least Developed Countries (Paris, 1981), Ethiopia presented a programme listing some 50 irrigation projects. Although these remained on paper, they caused considerable concern among the heavy users of fresh water downstream, namely Sudan and Egypt.

Egypt is by far the heaviest user of freshwater resources in the Nile basin and is totally dependent on the river. Indeed, as Herodotus put it long ago, Egypt itself is 'the gift of the Nile'. Egypt is virtually rainless, yet the Nile flood and carefully controlled irrigation combine with a long growing season to produce two and even three harvests a year. The first dam on the Nile was built at Memphis by Pharaoh Menes about 3000 BC, and the oldest preserved dam in the world, built only a little later, is at the Wadi el Garawi, 25 kilometres from Cairo. Basin irrigation was used in Egypt until the 19th century, when Mohammed Ali introduced perennial irrigation, and revolutionized Egyptian agriculture by constructing the Delta Barrage for water storage in the 1840s. During the colonial era, Britain controlled Egypt, Sudan and East Africa, and made arrangements for the disposal of the Nile's waters downstream. Modern irrigation was introduced at that time; the original Aswan Dam was completed in 1902, and was raised in height in 1933.

When the British colonial administration in Sudan conceived its own grandiose irrigation scheme for cotton production in Gezira, it became necessary to adjudicate the new claim on the Nile's waters. Accordingly, the Nile Waters Commission was set up in 1925, and in 1929 produced the Nile Waters

Agreement, which allowed Sudan a share of the water as long as it did not infringe upon Egypt's 'natural and historical rights'. The river's average annual flow having been calculated at 84 billion cubic metres, Egypt secured 50 billion to Sudan's 6. The agreement also committed Sudan and the British colonies in East Africa – Kenya, Uganda, Tanzania – not to construct any works on the lakes and rivers feeding the Nile that would affect its flow into Egypt. In 1959, Egypt and Sudan, now both independent states, reached an agreement to raise the former's share to 55.5 billion cubic metres and the latter's to 18.5. This had far-reaching consequences because it also confirmed plans for the building of the Aswan High Dam in Egypt, the Roseires Dam in Northern Sudan, and the Jonglei Canal in Southern Sudan.

Egypt's perennial concern has been to maintain a steady flow of water in the Nile through storage and to avoid intervention upstream. The Aswan High Dam has the capacity to store the entire Nile discharge for two successive years. Nature has not been obliging, and the Nile flow has undergone dramatic fluctuations. For example, during the early 1970s the mean annual discharge was close to 90 billion cubic metres, but fell to 72 billion cubic metres during 1977–87. Such variation underlines Egypt's vulnerability and the need for water control.

Although it uses three-quarters of the Nile flow, Egypt has a growing need for fresh water, given its rate (2.5%) of population growth. According to one calculation (Falkenmark, 1991, p. 88), Egypt will need twice as much water by the year 2025 in order to satisfy its population at the present demand level. Egypt proposes to meet future food requirements through desert reclamation; the Egyptian Land Master Plan proposes to reclaim 580,000 hectares through irrigation. Whether or not implementation of these plans can be accomplished within the limits of the 1959 Nile Waters Agreement is debatable. What is certain is that 'the implementation of Egypt's current desert reclamation plans will make it extremely hard politically for Egypt to consider reducing its allocation of 55.5 billion cubic meters in order to make room for water allocations for any upstream state' (Whittington & McClelland, 1992, p. 147).

There is potential also for cooperation or conflict over freshwater resources between Ethiopia and Somalia. Of the estimated 116 billion cubic metres of surface run-off carried annually by the 11 major river basins in Ethiopia, Somalia receives approximately 6%. The basins of the Juba and Shebeli Rivers, both of which rise in Ethiopia, contain most of Somalia's agricultural land, whose productivity depends directly upon the floodwaters of the two rivers. Any Ethiopian intervention in the upper reaches of the Shebeli or the tributaries of the Juba is bound to affect their flow and siltation rate and, in turn, affect agricultural production in southern Somalia. The Shebeli, Dawa and Ganale Rivers account for half of Ethiopia's power-generating potential. In 1988, Ethiopia completed the Melka Wakana hydroelectric project on the upper reaches of the Shebeli. While hydroelectric projects themselves do not consume water, they regulate river flow and create conditions conducive to irrigation. Somalia's plans to use the rivers for energy production could also

be adversely affected. A power plant is operating at Fanole on the Juba River
and is also used for irrigation. The unfinished giant Bardhere Dam was
designed to provide electricity for Mogadisho and Kisimayo, as well as to
control irrigation.

Ground water is a vital resource in the arid zone, where humans and animals depend on it for their lives. Pastoralists make use of wells in many parts
of the Horn, and borehole digging is an investment made by the state in the
pastoralist sector in order to facilitate the marketing of livestock. Inevitably,
the area around boreholes becomes devastated by the concentration of livestock; in Somalia borehole digging had to be stopped for this reason. The
location and pattern of aquifers in the Horn is not well known, except in
Sudan, where it is estimated that less than a third of the available ground
water is being used. The region's largest aquifer lies under the Nubian sandstone and covers an estimated 30% of the territories of Sudan and Egypt.

2.6 Marine Resources: Ports

Historically, the people of the Horn were fortunate in having access to the sea
and through it to other continents and cultures. Both the Red Sea and the
Indian Ocean served the region as links to the world outside Africa. The
Red Sea is the link between the Mediterranean and the Indian Ocean. A
narrow body of water 2,000 kilometres long, it measures 360 kilometres at its
widest point and 26 kilometres at the narrowest, at Bab el Mandab. It has
been a key sea-route since antiquity, lined with many ports, some of which
were initially described in the first century AD by an anonymous traveller in
the *Periplus of the Erythraean Sea* (Huntingford, 1980). Its importance
declined after the Europeans reached the Indian Ocean via the Cape in the
16th century and destroyed Arab sea power.

The opening of the Suez Canal in 1869 restored the Red Sea to world
importance. Its littoral became a prime bone of imperialist contention and led
to the scramble that divided the Horn among the European colonial powers.
Britain had the greatest interest because the Red Sea was a key link in its 'lifeline to India', and secured both ends of the sea-lane by laying claim to Egypt,
Aden and northern Somalia. Britain also claimed Sudan, where it built a
new harbour at Port Sudan. The Italians claimed the old port of Massawa,
and from there moved onto the northernmost part of the plateau to establish
the colony they named Eritrea. They also built a new port at Asab. The
French grabbed a slice of land in the Gulf of Tadjura, where they built the
port of Djibouti.

The imperialist occupation of the western Red Sea littoral prevented the
Abyssinians from gaining access to the coast, from where they had been
barred since the 7th century AD, when the Arab Muslim tide flooded the
Horn lowlands. An emerging regional power at the end of the 19th century,
Abyssinia checked Egyptian expansionism in the Horn, then inflicted a deci-
sive defeat on Italian imperialism in 1896 and retained its independence.

Furthermore, Abyssinia itself expanded prodigiously by conquest at the end of the 19th century, doubling its territory and population to create the Ethiopian Empire. The disadvantages of being landlocked were demonstrated during this period, when Britain and Italy imposed an arms embargo on Ethiopia in order to weaken it. The embargo failed because the French did not adhere to it, and made Djibouti a free port for Ethiopia, soon to be linked by rail to Addis Ababa.

Ethiopia's need for access to the sea intensified as the country became increasingly integrated into the world economy. Its quest for a port on the Red Sea was to be the main cause of a bloody conflict that lasted nearly three decades in the second half of the 20th century. After World War II, Ethiopia acquired both Massawa and Asab when it annexed Eritrea. The war of Eritrean independence that followed (see Chapter 5) left both Ethiopia and Eritrea in ruins, and the former once more landlocked. The new state of Eritrea has a coastline of 1,000 kilometres on the Red Sea and two major ports: Massawa, the best natural harbour in the Red Sea, and Asab on the Danakil coast. If they are not to remain perennial bones of contention, these ports, especially Asab, must be freely available to Ethiopia. By the same token, if they are to prove economic assets for Eritrea, they must handle Ethiopian and possibly Sudanese traffic. Indeed, Asab was declared a free port for Ethiopia after Eritrea's independence, and a customs union was arranged between these two neighbouring states.

Sudan has 700 kilometres of coast on the Red Sea. Suakin, whose existence was recorded as early as the 10th century AD, was the dominant port on this stretch of the coast and flourished after the opening of the Suez Canal in 1869. After the British colonial administration built Port Sudan, 58 kilometres to the north, in 1906, Suwakin declined and fell into decay. Located at the northeastern end of this vast country, Port Sudan serves areas that lie very far from it. Parts of Sudan would be better served by Massawa in Eritrea, for they are nearer to it than to Port Sudan. Djibouti occupies only 370 kilometres of the Red Sea coast. Obok was the old seaport here that was displaced by the one the French built on the opposite end of the Gulf of Tadjura. The port of Djibouti remains a major sea outlet for Ethiopia.

The Red Sea coast of northern Somalia runs for 1,000 kilometres. Two major ports of great age are found here. First mentioned in the 9th century, Zeila was the centre of the emirate of Adal and the principal port in the trade with Arabia and the Orient. It was overshadowed by Djibouti in recent times and fell into decline. Berbera, first mentioned in the 13th century, is still the outlet for the Harar plateau, the Ogaden, as well as northern Somalia, and is the principal port for the export of livestock to the Gulf region. Berbera has always served southeastern Ethiopia, whereas Zeila is ideally located to serve central Ethiopia. Indeed in the late 1940s, Ethiopia offered to exchange the valuable Haud region in the Ogaden for Zeila and a corridor leading to it. The offer was rejected by Britain, partly due to French objections stemming from fear that Djibouti's commercial position would be undermined.

The strategic significance of the Red Sea has international implications which made it the bone of contention among imperialist powers. In the post-colonial era, its importance was magnified when it became the route through which the Gulf oil was ferried to the international market. Naturally, the Arab states of the region are directly concerned, especially those that share the 5,300 kilometre coast of the Red Sea: Egypt, Jordan, Saudi Arabia and Yemen. It should be noted that three states in the Horn – Sudan, Somalia, Djibouti – are also members of the Arab League. Since the resurgence of Arab nationalism after World War II, and the rise of petroleum as a prominent economic and strategic factor on a global level, restoring Arab dominance in the Red Sea has been an oft-voiced Arab ambition. Mohammed Hassan Haikal, editor of Egypt's *Al Ahram*, in an article on 4 July 1969, declared the Red Sea ought to become an 'Arab lake'. Later in the same paper (27 October 1972), he proposed the establishment of 'an Arab naval command to control this waterway'.

Such Arab ambitions are anathema to other users of this strategic sea-lane. None of the world's great powers wants the Red Sea to become an Arab lake, and this consideration has guided their intervention in the affairs of the Horn. For example, they consistently supported Ethiopia in its struggle against Eritrean nationalism, for fear that an independent Eritrea would become dependent on the Arab world. This fear was nourished by the fact that several Arab states provided assistance to the Eritrean nationalist movement, and Arab publications often referred to Eritrea as part of the Arab world.

On the other hand, the two non-Arab states on the Red Sea coast – Ethiopia and Israel – became allies out of necessity and encouragement from their common patron, the United States. Israel became actively involved in Ethiopia's futile efforts to subdue Eritrean nationalism, as well as in the struggle of the Southern Sudanese to escape domination by the Arab North. Later, when the Soviet Union became Ethiopia's patron, these two countries issued a joint communique opposing 'the attempts by certain countries to establish their control over the Red Sea in violation of the legitimate rights of other peoples and states of the area and to the detriment of the interests of international navigation' (*Pravda*, 4 May 1977).

The Indian Ocean forms southern Somalia's border for 2,200 kilometres. The northern part of it has a number of small indentations, but no major harbours and no urban concentrations. The southern part, from Obbia to Juba, has very few indentations and no natural harbours, but contains a number of old ports, including Mogadisho – described by Ibn Battuta, who visited it in 1331 – Merca, Brava and Kisimayo. Along this coast are located the country's main urban centres and the bulk of its manufacturing capacity. In the past, none of these ports had berthing facilities, and ships had to anchor at sea to unload onto small craft. Good harbours were constructed later in Mogadisho, Merca and Kisimayo. Mogadisho is the principal port, normally handling half of all Somalia's traffic, and Kisimayo is important for the export of bananas and livestock. Merca has fallen into disuse. Kenya's Indian Ocean coast is 450 kilometres long, and contains a natural harbour at

Mombasa. Probably the most important port in East Africa and Kenya's second largest city, Mombasa is connected to Nairobi by rail, and also serves Uganda, which is reached by the railway. Mombasa is also an industrial centre. To the north of Mombasa lies Malindi, a minor port that has grown rapidly with the tourist industry.

2.7 Marine Resources: Fisheries

The Red Sea bed has a central trough reaching depths of more than 2,000 metres. From this trough the seabed rises abruptly to a terrace at a depth between 1,000 metres and 600 metres, then rises again to a continental shelf 300–400 metres deep. The width of this shelf varies along the length of the Red Sea, and it is this that determines the distribution of shallow-water marine habitats, because it is on this shelf that coral reefs are located. The shelf is narrow, not exceeding 10–15 kilometres, and the coral reef is limited. The Red Sea receives very little freshwater flow or nutrient-rich soil material, because no important river reaches it. Furthermore, being virtually a closed body of water, it has no nutrient-rich upwellings, except for minor ones from the Indian Ocean at Bab el Mandab. As a result, the Red Sea is relatively nutrient-poor, with low phytoplankton and zooplankton diversity. This means its capacity to support fish population is limited; a 1987 study concluded the Red Sea cannot support a major fishing industry (Head, 1987, p. 380).

Even so, the capacity of the Red Sea has hardly been tapped. What fishing is done is on a very small scale, especially in the countries of the Horn. Few people are involved, the means they use are elementary, catches are small and consumed locally. The bulk of the population in the countries of the Horn lives far from the coast, and the problems of transport and refrigeration are immense. As a result, saltwater fish consumption in the Horn is negligible: less than one kilogram per capita. Ethiopia's potential capacity (when it was still in control of the coast), augmented by its many lakes and rivers, was estimated at close to 100,000 tons per annum. Djibouti harvests only about one-eighth of its estimated annual potential of 4,000 tons. Egypt has the largest and most modern fishing fleet in the Red Sea, operating in the Gulf of Suez. Next in rank are Yemen and Saudi Arabia. These countries have much higher rates of fish consumption; indeed Egypt and Saudi Arabia, as well as Jordan and Israel, are major importers of fish products.

The continental shelf is widest at the northeastern coast of Somalia between 8° and 12° North, where it extends to 50 kilometres. An upwelling zone in that area also brings up nutrient-rich waters from the depths of the Indian Ocean. This is the richest area in fish resources, with commercially viable grounds for tuna, mackerel, herring, sardines, etc. Its potential is estimated up to 200,000 tons per annum. The Somali coast is dotted with small settlements where artisanal fishing is carried out with paddle-driven canoes and simple fishing methods. More than 20 fishing cooperatives were set up by the state in the 1970s and were provided with motorized boats and modern equipment which, however, did not stand the test of time. Commercial deep-sea trawling has been carried out by foreign companies under licence or in joint ventures with state agencies. Cold-storage facilities were available at Kisimayo, Las Koreh, Bolilog and Mogadisho. All told, about 10,000 tons were landed per annum in the 1980s, part of which was exported. Fishing occupied about 3,000 persons and was credited with contributing about 0.5% of the GDP in the early 1980s. Consumption per capita in Somalia was less than one kilogram.

The north Kenya banks of the Indian Ocean coast are considered relatively good fishing grounds with great potential for expansion. Commercial fishing with trawlers is carried out by foreigners. The coast has several estuaries in the mouths of rivers and creeks where prawn fishing is carried out on a commercial basis. Fishing occupies some 8,000 persons. Sea-fishing in Kenya is dwarfed by fish production from Lake Victoria, which accounts for 90% of all fish production and supports a domestic consumption of more than 3 kilograms per capita.

Because the main population concentrations have scant access to this food resource, saltwater fish does not feature in the diet of most communities in the Horn, and is consumed mainly along the coast. This is a cultural factor that will have to be overcome if this valuable source of protein is to be exploited in a region of perennial food scarcity. In Ethiopia, there is little consumption even of sweetwater fish that can be obtained from its many lakes and rivers. By contrast, such fish is widely consumed in the Lake Victoria basin in Kenya and in Southern Sudan. If the fish resources of the Red Sea are to be exploited in the future, protection measures will have to be taken now. The small, enclosed sea is highly vulnerable to pollution from the traffic that plies its waters and the activities that take place on its coast. In the future, the Red Sea will need to be protected from over-fishing and the uncontrolled use of methods like trawling. Such protection will have to involve all the littoral states in a regional approach.

2.8 Pastoralism*

The Horn is home to the largest remaining aggregation of traditional livestock producers in the world. Sudan, Somalia, Ethiopia and Kenya rank first, third, fifth and sixth respectively in the world in terms of pastoralist population size. Exact figures are not available, but it is thought that pastoralists comprise from 6% (Kenya) to 60% (Somalia) of the population. Vast expanses of land in the arid zone, which make up the largest part of each state's territory – 52% in Ethiopia, 66% in Sudan, 70% in Eritrea, 72% in

*This section incorporates data from the first part of Mohamed's Salih's contribution on 'Transboundary Pastoral Movements: Prospects for Cooperation in the Horn of Africa', presented at the PRIO/UNEP seminar in Sigtuna, Sweden, September 1991.

Kenya, 75% in Somalia, 100% in Djibouti – are pastoralist habitat. In the past few decades, the introduction of veterinary services and the stimulation of the market have resulted in a rapid rate of animal population increase. Two wet periods of exceptional rainfall (1919–34, 1950–65) added momentum to this trend.

In Sudan, it is estimated the total number of cattle multiplied 21 times between 1917 and 1977, camels 16 times, sheep 12 times and goats 8 times (Fouad Ibrahim, 1984, p. 125). Their numbers are estimated to have doubled between 1965 and 1986. Ethiopia's livestock population is believed to be the largest in Africa. Despite the presence of the tsetse fly in Kenya, that country is estimated to have over 10 million cattle and an equal number of flock animals. Tiny Djibouti boasts a million flock animals (see Table 2.2).

The pastoralist habitat should be considered a resource on the regional level, for several reasons. First of all, the pastoralists themselves often regard existing state borders as an obstacle to be circumvented, ignoring them whenever possible. Most pastoralists in the Horn had no experience with state structures until the colonial intrusion in their region. This was the time also when Abyssinia expanded hugely to become the Ethiopian Empire. Most state borders in the Horn were drawn through the lowland pastoralist habitat, and many pastoralist ethnic groups found themselves partitioned among several states. In the worst instance, the Somali were partitioned among no less than five state units: Ethiopia, French Djibouti, British Somaliland, Italian Somalia, British Kenya. Their Afar neighbours were divided between Ethiopia, Italian Eritrea and French Djibouti. The Beja in the north were bisected by the border separating Eritrea and Sudan, the Boran in the south found themselves on both sides of the Ethiopian border with Kenya, whereas the Nuer in the west were divided between Ethiopia and Sudan. Numerous smaller groups suffered a similar fate.

Not only did the geopolitical pattern imposed on the region violate the social and political integrity of pastoralist society, it also violated the imperative of the pastoralist mode of production for free movement of people with their herds. Furthermore, established patterns of trade were obstructed when a new centripetal pattern was imposed, designed to direct trade through the centre of the state in order to control and tax it. This new trade pattern

Table 2.2   *Livestock*

Country	Cattle	Camels	Sheep	Goats
Djibouti	190,000	62,000	470,000	507,000
Eritrea	1,550,000	69,000	1,510,000	1,400,000
Ethiopia	29,450,000	1,000,000	21,700,000	16,700,000
Kenya	11,000,000	815,000	5,500,000	7,438,000
Somalia	5,000,000	6,000,000	13,000,000	12,000,000
Sudan	21,751,000	2,956,000	22,870,000	16,448,000

Source: FAO, 1995. *Yearbook*, vol. 48, Rome.

worked against the interests of the pastoralists who are located on the periphery of the state and trade most profitably across the nearest border. The
borders established during the colonial period cut off many pastoralist groups
from their ethnic kinsmen, pastures and watering places, markets and places
of worship. It is scarcely surprising that the pastoralists should ignore these
borders as a matter of course.

It was during the colonial period also that cultivation was introduced in the
arid zone through irrigation, thereby beginning the process of depriving pastoralists of their most valuable pastures and permanent water places.
Pastoralist mobility was increasingly restricted also within each state by
provincial borders, tribal grazing boundaries, conservation schemes and quarantine zones. While the pastoralist habitat was shrinking as a result, livestock
numbers increased rapidly with the introduction of basic veterinary services.
The result was congestion, over-grazing and ecological degradation. Signs of
this ruinous trend were evident already in the 1940s, and were mentioned in
official reports concerning Northern Sudan, and in northern regions of
Kenya and Somalia.

The erosion of its material base undermined the economic viability of the
pastoralist economy. Needless to say, the pastoralist domain saw nothing of
the social and cultural advances of the colonial period. Pastoralist culture
remained unmodified and became increasingly remote from the modernizing
urban milieu that emerged during the colonial era. When that era ended,
pastoralism found itself on the primitive margin of a changing world. All the
trends that contributed to the decline of pastoralism accelerated in the post-
colonial period. Encroachment by cultivation into the pastoralist domain
advanced rapidly as new irrigation schemes were founded and old ones
expanded. Grazing land was also lost to national parks, game reserves and
conservation schemes. State investment in the pastoralist sector was negligible, and was designed simply to promote livestock marketing. This
encouraged livestock proliferation and accelerated land degradation.
Recurrent drought contributed to this trend, and pastoralists were repeatedly
devastated by famine. Today, large numbers of them are dependent on food
aid for their survival.

The plight of the pastoralists is a measure of their political impotence in the
post-colonial state, where they are normally ignored, unless they engage in
activities their rulers find threatening. Even in Somalia, where they constitute
a majority, their role in decision-making has been insignificant and their
interests ignored. This does not mean pastoralists are politically inactive.
Quite the contrary, they have been heavily involved in most political conflicts
in the region, and the nature of their involvement is another reason why pastoralism should be considered a regional concern (Markakis, 1993). Quite
simply, pastoralists often become involved in 'subversive' political activity:
that is, activity that threatens state security. They may launch their own movements or join movements launched by others, the common denominator
being resistance to the authority of the state. The list of confrontations in
which pastoralists have become involved is long, and includes the major

struggles in the region: the Eritrean revolution, the civil war in Sudan, Somali irredentism in Ethiopia, Djibouti and Kenya, and the clan wars that led to the disintegration of the Somali state. There are many less well-known episodes as well (see Markakis, 1987, 1993).

Most often pastoralists fight among themselves. This is not a novel phenomenon: fighting to establish rights over pastureland and water, and raiding to replenish depleted herds, are hallowed traditional practice. However, conflict among pastoralists in the Horn recently has taken on new, exaggerated dimensions. A shrinking resource base has provoked desperate struggles for survival, in which the very existence of groups is threatened. The struggle is waged with automatic weapons that claim large numbers of victims, and ignores customary rules that spare women and children. Such conflicts are often woven into larger confrontations, as the pastoralists seek allies and sources of weapons. One example is the arming of tribal militias by the Sudanese government to assist the military to fight the war in the South. Another is the clan struggle for state power in Somalia in the 1990s. Protracted violent conflict has accelerated the decline of pastoralism. Entire groups have been forced to leave their homelands and seek refuge across borders, where they spend years in sprawling refugee camps. Others have become internal refugees crowded into peri-urban slums. The economy of the largest pastoralist groups in the region has been destroyed and their homelands depopulated.

Pastoralists and their livestock represent a precious regional resource, because only they can make productive use of the vast arid expanse in the Horn. Today they are an endangered species, their survival a regional concern. The basic contradiction between pastoralism and the state means this problem must be approached on a regional basis. Only in this way can the fragmentation of pastoralist communities by state borders be mended, markets made accessible, trade revitalized, mobility restored, historical wrongs righted, and the integrity of an ancient way of life safeguarded.

2.9 Pastoralists on the Border*

There is hardly any major pastoral group which lives entirely within the boundaries of one state in the Horn of Africa. Pastoralists on the border exhibit all the features of divided and marginalized nations, many of which have established or joined some kind of liberation movement seeking separation or greater autonomy from the centralized states of the Horn. Despite this conflict there is a continuous flow of people and herds across borders, which can be attributed to at least four factors.

First, pastoralists lived in these territories long before the state boundaries of modern Africa were drawn and divided many nationalities between

*This is an abridged version of the second part of Mohamed Salih's contribution on 'Transboundary Pastoral Movements: Prospects for Cooperation in the Horn of Africa', presented at the PRIO/UNEP seminar in Sigtuna, Sweden, September 1991.

colonial states. The inherited colonial boundaries are main contributing factors to the civil wars in the Horn. Second, although the present boundaries are part of political reality for the countries of the region, the pastoralists consider them non-existent. Pastoralists depend for their livelihood on an environment characterized by seasonality, where seasonal movements through known migratory routes, established kinship networks, and long-standing traditional political alliances provide the bases for their way of life. Traditional management of resources reaches beyond state boundaries. The very existence of boundary controls negates any notion of pastoral survival based on seasonal movement in search of pasture. No matter which state now claims these resources, pastoralists believe they have always existed within their own 'ethnic' territory, which may extend over the borders of several of today's states and 'nations'.

Third, no pastoral community lives exclusively off the herd and its products. The recurrence of drought and epidemics has made pastoralists increasingly dependent on the import of grain from other regions. Their consumption patterns have changed, and they have become more dependent on local and international markets to sell animals and animal products in order to purchase grain and manufactured goods. Pastoralists on the border usually extend their domain of trade to markets and urban centres across state boundaries. Indeed, pastoralists consider such markets a natural extension of their traditional territory and part of wider social and economic networks essential for their survival. Fourth, the recurrence of drought, famine, and war has produced large movements of people seeking refuge across borders. Pastoralists in Sudan, Somalia, Ethiopia, Eritrea and Djibouti have been particularly affected by these calamities, and their presence is widely felt in the border towns of the region.

Although transboundary movements are essential for pastoral production and reproduction, state authorities are scornful of the disrespect shown by pastoralists for international boundaries. Administrators see transboundary movements as a hindrance to development, a cause of instability, and inconsistent with nation-building. Some pastoral groups comprise nations with a shared culture, language and widely perceived rights of access to a common territory (Mohamed Salih, 1990a). Consequently, they are easily suspected of divided loyalties and subversiveness, given their tendency to cross borders without paying attention to regulations and restrictions. On their part, pastoralists feel neglected by states that claim taxes but do not commit resources to enhance their welfare, nor delegate power so that pastoralists can run their own affairs. Dissatisfaction on both sides has led to violent confrontations.

Most border pastoralist groups have taken part in militant movements that challenged the authority of the state. The Beni Amer were among the first to enrol in the Eritrean nationalist movement. All Somali movements against Ethiopian rule earlier, and in the civil war in Somalia recently, relied on pastoralist manpower. Most pastoralist groups in Southern Sudan became involved in the rebellion against the North. The Afar in Ethiopia have their

Afar Liberation Front (AFL), and in Djibouti the Front pour la Restauration de l'Unité et de la Démocratie (FRUD). There is an obvious contradiction between the values enshrined in the modern, highly centralized, authoritarian state and the traditional values of political expression and leadership. In this situation, even genuine attempts by the state to reach out to the pastoralists through development projects have been frustrated by lack of contact points between the national and local levels of political action.

This contradiction is sharpened by state policies that curtail and suppress pastoralist aspirations for self-government. All states have responded to political demands for autonomy with intimidation and violence, turning the political arena into a battleground. The pastoralist 'regional' perspective has been challenged and shattered by national ruling groups struggling to maintain the post-colonial status quo. Any attempt to transform the situation from contradiction to cooperation will have to be orchestrated on a regional basis, because any development effort or administrative regulation of resource use in one state is likely to affect pastoralists in a neighbouring state. Likewise, resistance to state policies on one side of the border often affects brethren and kin on the other side as well.

2.10 Biological Diversity

'It is strange how little domestication of African animals (or of plants) has taken place, and the indigenous fauna and flora represent a biological reservoir of genetic material which it would be prudent to preserve for future need' (Morgan, 1973, p. 70). Morgan was referring here to East Africa, but the same applies to the Horn. The region's great diversity in climate, vegetation, temperature, elevation and terrain configuration, its vast, thinly populated lowlands, and its precious wetlands provide the setting for the development of a wide variety of fauna and flora species, many endemic to the region. A major attraction for tourism, the wildlife of the region has become its symbol. The Kenyan highlands and the Ethiopian–Eritrean plateau provide the habitat for an extensive collection of mammal and bird life. Ethiopia has 256 mammal varieties of wildlife and 832 varieties of birds. Kenya has 308 mammal species and 860 bird species; Sudan, 266 mammal species; Somalia, 173 mammal species and 645 bird species; and Djibouti, 327 bird species.

Wildlife throughout the region is endangered, and not a few species have already become extinct. Although hunting has been singled out as the main threat to wildlife and measures have been taken to control it, normally it is not the real culprit. However, in a region awash with automatic weapons, and large groups of heavily armed men waging political struggles from bases in the bush, conditions are not normal. Under these conditions, animals are likely to disappear completely from the vicinity. Nevertheless, the most relentless exterminator of wildlife is the impinging peaceful activity of man, not least the expansion of cultivation: 'as man moves in, animals must move out'

(Morgan, 1973, p. 73). Estimated wildlife habitat loss in the region ranges from 50% in Djibouti to 70% in Ethiopia and Sudan.

The Horn is equally rich in flora diversity. The Afromontane vegetation in Ethiopia numbers 4,000 plant species, while the total for the country is 6,283. The plateau has been identified as an important centre of primary and secondary genetic diversity in the world, and a plant genetic resource centre was established in Ethiopia in 1976 for the collection and preservation of germplasm. The vegetation in the area around the Imatong mountains in Southern Sudan represents half the varieties found in all of Sudan. The Imatong forests contain a wealth of genetic resources, such as wild strains of coffee and other plants not found elsewhere in the world. Northern Somalia has a variety of drought-resistant succulent plants. Numerous wetland ecologies are found in the Sudd, Makar and Kanamuka in Southern Sudan, the Rift lakes, the Danakil Depression, and in the permanent flood zones of the lower Shebeli and the Awash Rivers.

This wealth is similarly endangered by human activity, as the expansion of cultivation results in denudation of land. The Horn has already lost most of its forest cover and an unknown number of flora species along with it. In a region of scarce energy and fuel resources, charcoal-making for use in urban areas and fuelwood collection in rural areas is taking a steady toll. Irrigation changes the ecological setting in which native plants developed, and leads to loss of species. Local crop varieties and their weedy relatives are an important genetic resource that is rapidly disappearing as imported varieties replace local ones, and new cultivation methods eliminate the weeds. This process accelerates during periods of famine when people are unable to keep their own seeds for later planting. Native food-producing strains, like Sudan's *dura*, are fast being replaced with foreign 'improved' varieties, and the traditional sorghums and millets are being replaced by maize and wheat. Native cotton has long since been supplanted by 'improved' varieties. In Kenya, hybrid maize is widely used, artificial insemination is promoted in livestock breeding, and afforestation programmes use improved germplasm.

Loss of biological diversity poses a double risk for the region. Biotechnology produces new crop plants with higher yields and resistance to pests. Their major characteristic is uniformity and consequent loss of variety in plant species. Thus, thousands of rice species across the globe are gradually replaced with only one species. Africans traditionally relied on risk-aversion methods based on multiple cropping of a variety of food-producing plants adapted to local conditions and less vulnerable to the whims of nature. Reliance on a few imported species limits the flexibility that provides some insurance against failure. Second, new improved breeds are developed abroad and must be imported into Africa. In the dawning age of bio-technology, these products come under patent and licence protection, the cost of which will be borne by the cultivator in Africa. Ironically, most of the genetic raw material of bio-technology is provided by the Third World in the form of wild plants, without compensation.

Plant genetic resources found in the developing world are considered 'a common heritage' of mankind, available to researchers and enterprises in the developed world. Rising concern in the developing world about this laissez-faire practice was expressed at the United Nations Conference on Environment and Development (1992, Rio de Janeiro). One product of that conference was the drafting of the Convention on Biological Diversity, signed initially by most countries of the world, but ratified by only a few thus far. Article 15 of the Convention recognizes the 'sovereign rights of states over their natural resources', and gives state governments 'authority to determine access to genetic resources within their territories' on 'mutually agreed terms'. The protection of these rights requires technological and organizational capacity sadly lacking in the Horn of Africa.

3

Country Resources

This chapter describes the resource endowment of the countries in the Horn of Africa. It focuses on land and water, because cultivation sustains the overwhelming majority of the people and produces nearly all the exports. Particular attention is paid to the modes of production in the agricultural sector, whose development is not simply the outcome of man's interaction with nature, but is shaped to a considerable extent by the social and political forces that impinge on it. The region appears poorly endowed with mineral wealth, though this is partly due to poor knowledge of its resources. Recent discoveries of petroleum in Sudan and natural gas in Ethiopia have spurred exploration. The same holds true for the energy sector, where traditional sources predominate (Table 3.1). Yet, the region has abundant potential resources for hydroelectric and thermal energy production.

Table 3.1   *Fuel and Water Consumption*

Country	Traditional fuel as % of total energy use	Commercial energy consumption per capita (kg of coal equivalent) Africa – continent average 432	Water consumption per capita (cu. m)
Djibouti	–	512	28
Eritrea	–	–	–
Ethiopia	91	30	48
Kenya	79	114	48
Somalia	85	28	–
Sudan	81	63	–

Sources: World Resources Institute, 1994. *World Resources 1992–1993, Washington*; UN, 1995. *Statistical Yearbook*, 40th Issue, New York.

3.1 Ethiopia

Ethiopia is the third largest country in Africa in population – after Nigeria and Egypt – and by far the most populous country in the Horn. More than half of Ethiopia's land lies in the arid lowland region and is utilized for grazing by mobile pastoralists. Less than 25% of the total land area is devoted to cultivation and grazing by sedentary peasants, a little less than 4% is occupied by forests, 2.5% is woodland and nearly 16% is sand or exposed rock. Lakes cover 7,000 square kilometres. This pattern of resource distribution determines the distribution pattern for population. The bulk of Ethiopia's inhabitants are congregated on the highland region, at altitudes of 1,000–3,000 metres. Constituting 47% of the total area, the highlands accommodate 74% of the population. A narrow belt of very high population density stretches north to south along the centre of the country, from Gemu Gofa to the border with Eritrea. This belt contains 65% of all cultivated land and 61% of the rural population (*National Atlas of Ethiopia*, 1988). Population density in this belt ranges from 60 to 300 per square kilometre, while the average density for the entire country is 34 per square kilometre.

Ethiopia is one of the least urbanized countries in Africa: with nearly 90% of its people in the rural sector, it lags behind all its neighbours in the Horn in this respect. Having such a large proportion of people in the rural subsistence sector, the fruits of whose labour appear only nominally in national statistics, limits the value of statistical indicators offered in official reports. According to such indicators, and following the catastrophes of war and famine throughout the 1980s, Ethiopia had the lowest per capita income in Africa in the early 1990s. On the basis of most other indicators as well, it ranked as the continent's poorest country.

It was not always so. The development of cultivation on the northern plateau among the Abyssinians was linked to the experience of North Africa and the Middle East, rather than sub-Saharan Africa. Ample precipitation and the use of the ox-drawn, steel-tipped plough made intensive cultivation possible, raising the output per unit of land and of labour above that reached by traditional hoe and shifting cultivation methods elsewhere in Africa. High yields and high population densities raised the value of land and made control of it, rather than control of labour as elsewhere in Africa, the fulcrum of the stratified societies and centralized states that appeared on the northern plateau.

The Abyssinian peasantry cultivate the reddish volcanic soils of the highlands intensively with multiple cropping, crop rotation and fallowing. The staple item of the Abyssinian diet is *teff*, a locally developed grain with high calcium and iron content. *Teff* has one of the lowest yields among cereal crops – 8.8 quintals (one quintal = 100 kilograms) per hectare – compared with maize and sorghum – 15 and 14 quintals respectively. *Teff* and barley are cool-weather crops, while maize, sorghum and millet are drought-resistant, warm-weather crops. Root crops (sweet potatoes, white potatoes, onions), legumes (horsebeans, lentils, chickpeas) and vegetables (peppers, okra, *berbere*) are also grown. The Abyssinians also keep a large number of cattle, goats and sheep, but use neither manure fertilizer nor irrigation in any form. Cow dung is used for fuel.

In a real sense, agriculture in northern Ethiopia is the victim of its own earlier success. It not only supported a stratified society with a sizeable aristocracy and clergy, but also produced the highest population density in the region, next to the riverine zone of Northern Sudan. Population increase led to continuous fragmentation of cultivated land, and the majority of the peasants worked parcels of land that kept diminishing in size, reducing simultaneously the efficiency of peasant labour. The first landholding survey carried out in three northern provinces in the 1960s showed two-thirds of the holdings were less than one hectare in size, and close to one-half were less than half a hectare (Markakis, 1974, p. 81). The situation in this respect can only have changed for the worse since then.

Increasing human and animal population pressure on land forced the peasantry to use methods that led to land degradation and ultimate disaster. Forest and woodland were cleared for cultivation. The use of wood for fuel accelerated the process of land denudation, as did the use of scrub and bush for fencing and animal enclosures. From the 17th century onwards, foreign travellers noted the speed with which wooded cover was being eliminated, and the lack of any care for its replacement (Pankhurst, 1968, pp. 243–247). Urbanization, a recent development in the highlands, created an incessant demand for charcoal and firewood, and wooded areas near towns were quickly stripped bare. Today there is hardly any forest cover on the northern highlands, where the mountain flanks have been eroded down to their stony spines. The sole attempt in the past to restore wooded cover was undertaken at the end of the last century under Emperor Menelik, to alleviate a severe wood shortage in the environs of his new capital, Addis Ababa. Two species of eucalyptus trees imported from Australia fared exceedingly well, and spread to all parts of the highlands to become the most important source of firewood and construction material.

The line of cultivation has descended steadily down the flanks of the plateau, onto less suitable land and poor soils. Steep slopes and ravines on the plateau came under the plough, exposing the land to rapid water and wind erosion. Land denudation decreased water retention and accelerated water run-off, which in turn increased erosion and lowered land productivity. Good practices like fallowing and terracing had to be restricted or abandoned, and even crop rotation had to be curtailed in order to satisfy the demand for staple food crops. As the sources of firewood moved farther away, animal dung and farm residue came into use as energy sources, depriving the land of fertilizer and animals of fodder.

Land erosion and falling productivity are the twin afflictions of agriculture in the northern highlands. Astounding figures are cited concerning the quantities of precious topsoil lost through erosion, and apocalyptic images are drawn of what the highlands will look like in the foreseeable future if this ruinous process continues. Two thousand tons of soil per square kilometre each year is the figure cited in *The Ethiopian Experience*, the Ethiopian government's submission to the UN Conference on Desertification (Nairobi, 1987). The source of this estimate is a Swedish agronomist who did research in Ethiopia in the early 1980s (Hurni, 1987). While it may seem quite fantastic, one need only see the rich reddish-brown colour of the waters rushing down the rivers and streams of the highlands during the rainy season, to realize that a good deal of soil is indeed carried along. One estimate puts the share of soil in the volume of water at 10–15% (FAO, 1986a).

Conventional wisdom blames the Abyssinian peasant for 'centuries of misuse' of the environment, as if peasants were ever free agents to select the optimal use for their land. Such conventional wisdom ignores the crushing burden of a feudal system borne by the peasantry until the second half of the 20th century. A multiple taxation system robbed peasants of the surplus of their labour and stifled desire to raise productivity (Mesfin Wolde Mariam, 1986). The absence of even an elementary transportation system was another disincentive, for it precluded the development of local markets. Continuous land fragmentation also discouraged investment in permanent modifications that could not be retained when the land was divided. While the Abyssinian peasantry had secure customary rights to land as long as it paid taxes, the high value of land made it the object of intense competition, and the peasantry was entangled in perennial litigation and in thrall to a proverbially corrupt judiciary.

Disequilibrium between population and resources was clearly a prime factor in the spectacular expansion southwards of the Abyssinian kingdom at the end of the last century, which brought the southern half of the plateau and the surrounding lowlands into the domain of what was thereafter called the Ethiopian Empire. Far less densely populated, the southern half of the plateau retained its flora and fauna in almost pristine condition. The land was cultivated with the hoe, and *ensete* (false banana) was the staple food for some of the ethnic groups of the area. After the conquest, the Abyssinians divided the land into three portions, two of which were claimed by the state. State land was distributed to northerners who flocked to the south to take advantage of cheap land and the labour of the indigenous people who were subjected to quasi-serfdom. Since then, northerners have been drifting south, relieving some of the pressure on their home region.

The Abyssinian expansion spilled onto the lowlands surrounding the plateau, bringing into the Ethiopian Empire many pastoralists who were frequently cut off from their kinsmen by the border. The arid lowlands are a strikingly different environment from the highlands, and few highlanders sought to settle there. Nevertheless, the pastoralists were to prove no end of trouble for the Ethiopian state. The size of the pastoralist population in Ethiopia is not known, and estimates vary from 8% to 12% of the total. Lowland pastoralists are estimated to hold a sizeable share of the country's livestock population: 21% of cattle, 25% of sheep, 75% of goats, and one million camels (UNDP, 1984).

The feudal system collapsed in 1974 in the wake of a devastating famine. Then, under a radical military regime, Ethiopia attempted to leap from the feudal age to a Communist utopia. Land was nationalized, the landlords were expropriated, and the working peasants were given usufruct rights over the land they tilled. Selling or renting land was forbidden, and hiring of labour was outlawed. The peasants were grouped into associations, presumably to run their own affairs. This all seemed a promising start, but it proved a crushing failure. Caught in a maelstrom of civil wars, the regime levied multiple taxation on the peasants to raise funds, and forced them to sell their produce to the state at fixed prices, in order to secure cheap food for its armies and the townspeople. Peasants were forcibly conscripted to serve as cannon fodder in the regime's wars, and their associations were turned into adjuncts of the state in order to implement its policies. The result was falling productivity and a worsening food deficit. The situation was compounded by persistent drought that caused the worst famine in Ethiopia's recorded history in the early 1980s. Despairing of the peasantry, the regime tried to promote state farms, which absorbed nearly all the state investment in agriculture, with little to show for it. Neither this regime nor its predecessor devoted any effort or capital to increase food production in the subsistence sector.

Coffee is the mainstay of Ethiopia's commercial agriculture. Coffee grows wild in the southwest – where it is said to have acquired its name from Kaffa, the coffee-producing region – and in the southeast. Commercial cultivation began when the inland port of Gambela in the southwest began operating as a free port under an agreement between Ethiopia and the British colonial administration of Sudan. For a long time, coffee was picked from the wild and roughly processed; consequently, it commanded low prices. Plantation production has evolved slowly, and processing is still inadequate. Ethiopia exports about half its coffee production, and only one-quarter of the exported coffee is clean washed; by contrast, about 90% of Kenya's exported coffee is clean washed. Coffee earns about 60% of export value. *Khat*, a mildly narcotic plant of increasing demand in the Horn and across the Red Sea, particularly in Yemen, has developed into a major commercial product, distributed mostly through the informal market. Hides, another staple export item, are second to coffee in value. Sugar production was developed in the Upper Awash valley by a Dutch company in the late 1950s, and cotton production expanded rapidly in the Middle Awash valley in the 1960s.

Water

Ethiopia is the water tower of the Horn, supplying Somalia, Sudan and distant Egypt. Ethiopia itself makes slight use of this resource: only an estimated 3% of the run-off is retained within the country. In Ethiopia, 85% of the population has no access to clean domestic and potable water, and little use is made of water for human waste disposal or industrial purposes. A high incidence of diarrhoeal and enteric diseases, as well as malaria and worm infections, is the result. Only a beginning has been made towards harnessing the country's huge hydroelectric potential, and mainly the Awash River has been utilized. The Awash basin in the Rift Valley covers an area of 120,000 square kilometres of rainless, but intermittently flooded and waterlogged, malaria-infested land. The first dam on the river was built during the Italian occupation (1936–41) outside Addis Ababa. In the 1960s, three more dams for power generation were built, opening the region to irrigated cultivation. Sugar, cotton, vegetables and fruit are grown on about 70,000 hectares.

All told, only 100,000 hectares of irrigated land are cultivated in Ethiopia, whereas the estimated potential for the Awash basin alone is 175,000 hectares. A study carried out by the United States Bureau of Reclamation during 1958–63 proposed 33 irrigation and hydroelectric power generation projects for the Blue Nile basin. The total area of proposed irrigation projects was 434,000 hectares, with a water requirement of 6 billion cubic metres (Guariso & Whittington, 1987). Two small power plants have been constructed on the Blue Nile: one just below the river's outlet from Lake Tana, and a larger one on the Fincha River, a tributary that drains Lake Fincha on the central plateau. In 1988, the Melka Wakena hydroelectric project on the upper Shebeli was completed, and two others were under construction at the time: one on the Gibe River at Gilgel, and another at Amerti.

Minerals and Energy

As the *National Atlas of Ethiopia* (1988, p. 55) candidly put it, 'Ethiopia was not exposed to a long period of colonial rule and therefore it did not inherit an accurate data base concerning its mineral resources'. There is no definite knowledge of the country's mineral endowment. Gold has been plentiful throughout its history. It was found in alluvial deposits along the Blue Nile below its junction with the Didesa River, and in western Ethiopia – Beni Shangul, Wallega, Kaffa, Sidamo. Gold is still produced by a state-owned enterprise at Adola in Sidamo Province, averaging more than 500 kilograms annually. Since private production carried out by individuals evades state control, overall production is much higher than that. A small amount of platinum is produced at Yudbo in Wallega Province. Other minerals produced are manganese ore, potash, feldspar, raw material for cement and silica sand. Copper, zinc and nickel deposits have been identified, as well as phosphates, sodium chloride and tantalum. According to the *Statistical Abstract* for 1992, mining contributed less than 1% of the estimated GDP (Ethiopian Transitional Government, 1993).

Lying opposite the Gulf, the world's most productive region for petroleum, the Horn is thought likely to have petroleum in its own subsoil. In Ethiopia, the Ogaden basin is considered the most promising petroliferous area, and explorations by foreign concessionary companies have been going on there intermittently since the 1950s. Similar explorations were undertaken offshore in the Red Sea, when Ethiopia was still in control of the Red Sea coast. In the 1980s, oil traces were discovered in wells dug in the southern Ogaden and southern Bale, and large deposits of natural gas were discovered in the southern Ogaden. Plans were under way in the early 1990s for gas production that could ease Ethiopia's crippling oil import bill, which reached 22% of national expenditure in the 1980s. Ethiopia imports crude oil for refining at a plant which was built in Asab on the Red Sea in 1965, and which is now under Eritrean control.

Even by African standards, Ethiopia's energy production is woefully inadequate. Electricity production per 1,000 inhabitants was 9 kilowatts in the early 1990s, while the average for the Least Developed Countries was 107 kilowatts. Traditional sources of energy like firewood, charcoal, animal dung and farm residue accounted for 91% of energy consumption, with devastating effects on the environment. To satisfy the energy needs of a few towns with a population of 10,000 and above, more than 100,000 hectares of woodland were being cut down every year (*National Atlas of Ethiopia*, 1988, p. 52). Deforestation in the 1980s was estimated to have stripped 88,000 square kilometres, while reforestation restored only 6,000 square kilometres (Pearce & Turner, 1990). Ethiopia has great geothermal resources, yet to be exploited, especially in the Rift Valley.

3.2 Eritrea

The plateau in Eritrea is an extension of the highlands, with an average altitude of 2,000–2,500 metres and one peak, Amba Soira, soaring over 3,000 metres. The northernmost spur of the highlands narrows as it descends to the Sudanese border. The western flank of the central plateau is a wide, broken plain that descends gradually towards the Nile basin. To the east, the plateau drops abruptly to a narrow coastal strip, nowhere more than 80 kilometres wide. South of the Gulf of Zula, the coastal zone widens to include the Danakil Plain, a barren region where the depression known as the Kobar Sink falls nearly 100 metres below sea level. In the Dahlak archipelago off the Red Sea coast, there is a cluster of some 100 small coral islands, few of which are inhabited. In the Hanish archipelago towards the southern end of the Red Sea are a few more uninhabited islands, possession of which became a bone of contention between Eritrea and Yemen in the mid-1990s.

The Eritrean portion of the plateau is drained by five major rivers, all of which flow into Sudan. Two of them – the Tekeze and the Mereb – form the border with Ethiopia. None of the other three rivers – Gash, Barka, Anseba – have a continuous flow throughout their course during the year. Several seasonal streams drain the plateau on the east and some manage to reach the Red Sea. The Eritrean section of the plateau receives the lowest precipitation on the highlands, varying between 400 and 800 mm annually, and so unreliable that 95% of Eritrea is classified drought-prone. The eastern escarpment between 700 to 2,000 metres, however, is a high precipitation zone with an average of over 1,000 mm. The coast has very high temperatures that reach record peaks in the Danakil Depression, and low precipitation of 150 mm or less. The western plain receives less than 400 mm.

Vegetation cover varies widely. The few available records indicate a rapid elimination of trees during this century. Eritrea now has little true forest cover and only sparse woodland. Acacia trees and shrubs are the predominant species on the plateau. The western plain is savannah land, while seasonal, sparse grasses are found in the arid eastern lowlands. Scarcity and unreliability of rainfall make cultivation a hazardous enterprise in Eritrea. Italian and British rainfall records since 1938 show that less than one-tenth of the land receives more than 500 mm and can be considered arable. Actual figures of land use at that time put the size of cultivated land on the highlands at little more than 4% of the total, and less than 3% for Eritrea as a whole. Three-quarters of the whole has been classified as grazing land (United Nations, 1950, pp. 26–27). The first assessment carried out by the Ministry of Agriculture of independent Eritrea put the total area of cultivated land in 1993 at a little less than 400,000 hectares (Hawando, 1994).

What was written above about the state of agriculture on the northern plateau of Ethiopia applies to Eritrea as well; only the situation here is worse. Intensive cultivation is carried out only on the plateau, which is the physical as well as demographic extension of the Abyssinian tableland, and is inhabited by Tigray-speaking, Christian peasantry. Sorghum, millet, barley and *teff* are the staple crops. The central plateau, the granary of Eritrea, is densely populated, with up to 300 persons per square kilometre. Although it occupies only about one-quarter of the territory of Eritrea, it is home to more than half its population. Cultivation on the Eritrean plateau probably has a longer history than elsewhere in the highlands. This section of the plateau also has the worst incidence of erosion by rain and wind, and a large part of its soil is highly degraded. As a result, crop yields in Eritrea are only half of those elsewhere in the highlands.

Despite great hopes and considerable effort, Italian colonial rule did not produce an export crop in Eritrea, although it did promote vegetable, fruit, tobacco, coffee and cotton cultivation. Foreign capital, mainly Italian, returned to Eritrea in the 1950s and 1960s to invest in cotton, fruit and vegetables. In those days, Eritrea was exporting fresh fruit and vegetables to the Middle East. However, no investment was made in food production, and Eritrea remained dependent on grain imports from Ethiopia and Sudan.

Half of Eritrea's population are thinly spread over the northernmost spur, the western plain and the eastern seaboard. Many people cling to the flanks of the plateau, pursuing precarious cultivation and raising animals. Traditional livestock raising is the dominant economic activity in the lowlands, although most people try to grow crops where soil and weather conditions permit. Some cotton also is grown in the lowlands. In normal times, livestock is a valuable resource and has the potential to play an important role in Eritrea's foreign trade. Eritreans raise cattle, goats and sheep, and there is a market for these across the Red Sea. In the 1950s and 1960s, Eritrean livestock was exported there, while hides and skins were exported to other parts of the world.

Eritrea won its independence after a violent struggle that lasted nearly three decades. During that period, especially the late 1970s and the 1980s, the country was laid waste and its economic assets were heavily damaged. Land denudation was accelerated by the requirements of an Ethiopian army that numbered 150,000 at its peak. Half of this number was concentrated in and around Asmara, the capital city of Eritrea. The army's need for fuel and construction material to build shelters, trenches and fences resulted in the elimination of all wooded cover in the vicinity. Throughout Eritrea, napalm, defoliants and other means were used to eliminate vegetation for security reasons. One observer commented: 'For the Ethiopian trooper a green valley surrounded by bare mountains is a guerrilla haven which should be cleared for better security' (Zeremariam Fre, 1991, p. 136).

Land degradation occurred in areas where people were forced to congregate due to the war. This was around major urban centres where people sought refuge from the violence that was sweeping the countryside, and around the 'security hamlets', where people were concentrated in order to minimize contact with the Eritrean nationalist guerrilla army. People were forbidden to move beyond a limited distance from these hamlets. Dense concentrations of humans and livestock, in addition to the presence of Ethiopian military personnel, put an intolerable strain on the land and water resources of these areas. By contrast, other areas were depopulated when they became 'free fire zones' – they were sown with mines and anything that moved there could be shot at – or because people fled to seek refuge elsewhere. In 1993, the Food and Agriculture Organization (FAO, 1993b) warned that 'the overall environmental situation in Eritrea has reached a critical stage'.

Pastoralists suffered more than most, especially in the western plain of Eritrea, where the nationalist struggle began in the early 1960s. The Beni Amer pastoralists provided its first fighters, and their homeland was the first to be devastated by Ethiopian army pacification campaigns. The first of these occurred in 1967, and the Beni Amer were the first Eritrean refugees in eastern Sudan, where they found themselves in a familiar environment among their Beja kinsmen. Having lost their animals, many Beni Amer turned to cultivation, and many appear to have settled in Sudan permanently. Pastoralist movement patterns were completely disrupted by the division of Eritrea into government-held and rebel-held areas, and herd movement became increasingly curtailed, with disastrous consequences for land and animals. Drought also forced pastoralists to move into Sudan, where their animals were sold at very low prices. According to an estimate by the government of independent Eritrea, 70% of the cattle were lost in the preceding decade (Eritrean Provisional Government, 1993).

Eritrea's infrastructure was thoroughly demolished during the war. The Italians had provided their colony with a superior rail and road transportation system, as well as a second port on the Red Sea. The port of Massawa was linked by rail to Asmara, and lines from there reached to Agordat in the west and Keren in the north. All-weather roads were constructed on the plateau and linked it to the lowlands. The long war took a terrible toll: the 306 km railway was dismantled, the port of Massawa was disabled by bombing from the air, and the road system was literally worn out from lack of maintenance.

Water

With five rivers draining the plateau, Eritrea has considerable freshwater potential. However, the fact that three of its rivers do not maintain full flow throughout the year limits their usefulness. On the eastern plain, the alluvial soils borne annually by the torrents coming down from the plateau are intensively cultivated. The Italians introduced modern irrigation on the Gash River at Tessenei to grow cotton. After World War II, Italian capital returned to Eritrea, and cultivation of fruit, vegetables and cotton expanded. A small dam was built at Zula near Massawa in 1960 with a capacity to irrigate 4,000 hectares. All these were lost in the war. Estimated surface run-off is nearly two billion cubic metres, and an estimate of the country's irrigation potential is 0.6 million hectares (Eritrean Provisional Government, 1993).

Minerals and Energy

Historically, Eritrea was a major producer of gold. The Italians worked several gold mines in the late 1930s on the plateau and the western plain, but most of these were discontinued after World War II. Salt mined in the Danakil Depression is another traditional product. Deposits of copper were developed by a Japanese company in the 1970s, and the first shipment abroad took place in 1974, but work was halted soon afterwards due to the war. Iron, zinc, nickel, asbestos, potash, and feldspar deposits are known to exist. The Italians explored for oil and carried out drilling in the Dahlak islands, without result. Exploration was continued by foreign companies under Ethiopian rule, and oil traces were found in the Dahlak islands. Eritrea now owns the oil refinery at Asab, upon which Ethiopia also depends. Hydroelectric power generation does not exist in Eritrea, but there is potential for it, especially from the Tekeze and Gash Rivers. The potential for geothermal power generation is considered good, particularly in the Danakil region.

3.3 Sudan

The bulk of Sudan's territory – nearly two-thirds of it – is classified arid according to conventional criteria. Unimproved grazing was the traditional pattern of land use in this area, and the bulk of its inhabitants are pastoralists who make up an estimated 12–15% of the total. Cultivation has a long history in some parts of this vast country. Historically, the riverine region of Northern Sudan had the most advanced agriculture in the Horn. Successive waves of Arab immigrants from Upper Egypt settled here and assimilated the indigenous population. Various irrigation methods were used to cultivate narrow strips of land that crossed the desert like green ribbons. These included the *shaduf* (counter-weighted lever), the *saqia* water wheel and basin irrigation. Main crops were cereals, beans, vegetables, fruit, dates and cotton. Intensive cultivation was also practised in eastern Sudan in the alluvial soils of the inland deltas formed by the Gash and Barka rivers.

An immensely larger area devoted to agriculture is the savannah region of central Sudan, where rainfed cultivation is possible. This is a vast area between latitudes 9°–15° North, representing about one-third of the total area of the country and over half of Sudan's arable land, and is home to nearly half of its population. Rainfall here increases southwards, ranging between 300 mm in the northern fringe to 800 mm in the south. Shifting cultivation was practised here in an ecologically balanced cycle that combined crop raising with the production of gum arabic, a major export product. The cycle began with the burning of acacia scrub and the cultivation of crops for four to ten years. The land was then left fallow, to be invaded by acacia trees. After eight years the trees were tapped for gum arabic for six to eight years. The acacia garden also produced firewood, building materials and fodder. When the trees began to decline, they were burned and the cycle began again. This system made extensive use of land, and each family had several fields. *Dura* (sorghum), the staple food crop in Sudan, from which beer is also produced, was grown in many varieties to suit ecological variations. *Dukhn*, a quick-maturing millet that can survive in the drier northern fringes of the savannah region, is another staple crop. Sesame, groundnuts and cotton are also grown.

Change in the agricultural sector was initiated in Northern Sudan during the Egyptian occupation (1821–81), when cotton cultivation was introduced in the inland deltas of the Gash and Barka, indigo and sugar were produced, and riverine cultivation intensified with the proliferation of *shaqia* waterwheels and slave labour. Northern Sudanese agriculture was revolutionized during the colonial period, when cotton became king. The immense Gezira scheme on the Blue Nile covered nearly 400,000 hectares. Smaller schemes were established in the inland deltas of the Gash River in Kassala and the Barka River at Tokar. At the same time, mechanical pumps were introduced to expand the riverine cultivation far beyond the narrow limits of traditional irrigation methods. Towards the end of the colonial period, the so-called Mechanized Crop Production Schemes introduced tractor cultivation in the eastern part of the savannah zone, where shifting cultivation had been the rule till then. Cotton was the principal crop throughout, accounting for 62.4% of all export value by 1956.

After independence, the main features of the Sudanese economy remained the same, with emphasis on export-crop cultivation and cotton at the top of the list, and heavy reliance on irrigation. The Gezira scheme soon doubled in size, and new schemes were constructed at Khasm el Girba, Rahad, Roseires and Kenana. More than half the irrigated land was under cotton. The exceptionally wet period of the 1950s led to the rapid extension of mechanized farming on the rainfed lands of the central savannah zone. The vastness and flatness of this area made it suitable for large-scale mechanized farming, and private capital was attracted by the promise of quick profit. The savannah was inhabited by subsistence cultivators and pastoralists with no secure title to the land, and they were easily dispossessed by state fiat. The Mechanized Farming Corporation surveyed and demarcated the land for schemes that were leased to individuals and groups. Initial payment and annual leasing fees, as well as the requirement of owning at least one tractor and a disk harrow, ruled out peasant participation.

Capital was made available to Sudanese entrepreneurs by the state and international funding institutions. The size of the land allotted to each farmer expanded continuously and, in time, became enormous. The limit for an individual scheme was between 800 and 1,000 hectares, but there was no limit to how many schemes an individual or a group could operate. The operators of these schemes were no ordinary farmers, but urban entrepreneurs and persons with political influence. Lured by the mirage of turning Sudan into the 'breadbasket' of the Middle East, capital from the Gulf states invested enormous sums in Sudanese agriculture, often in partnership with Western multinational firms which provided technological and managerial expertise. The scale of these ventures was immense. For example, the Kenana sugar scheme, managed by Lonrho, was said to be the biggest in the world. According to official figures, 5.5 million hectares were mechanically cultivated in 1991.

The initial losers in the development of commercial agriculture were the pastoralists. The first to be affected were the livestock herders in the Gezira plain, who had constituted 90% of its population before the scheme began. By the time the scheme was fully developed, the pastoralist population had been reduced to one-tenth. The Beja pastoralists in the Tokar and Kassala areas of eastern Sudan also lost their pastures in the inland deltas of the Barka and Gash, while the Lahawin were dispossessed by the Khasm el Girba project. Not only prime pastureland was lost, but also the land on which the pastoralists planted crops for themselves. In some schemes, pastoralists were offered plots of land to cultivate cash crops on a sharecropping basis. Initially, not a few took up the offer, only to find out quickly that it meant giving up their herds. Subsequently, most of them leased their plots to third parties, and afterwards took little interest in the schemes. The inroads made by mechanized farming in the savannah had similar dire effects on the pastoralists of western Sudan.

Usually, the launching of irrigation or mechanized farming schemes begins with the clearing out of all trees and vegetation cover. The Lahawin remember 1958, when work on the Khasm el Girba Dam began, as 'the year of cutting trees' (Berhane Woldegabriel, 1991, p. 106). Wildlife disappears soon afterwards. The interposition of immense cultivated areas in areas of livestock movement becomes a source of conflict between pastoralists and farmers. The pastoralists find their movements obstructed, and feel harried by the need to keep their herds from straying into cultivated land to avoid incurring heavy fines. They find the waterholes contested by farmers who see the herds as a threat to their crops, and who have no need to enter into exchange arrangements with the herders, as subsistence cultivators did in the past. The intrusion of cultivation forces pastoralists to compete among themselves for increasingly scarce resources. Indeed, conflict among pastoralists for pastures and water is far more frequent than conflict between pastoralists and cultivators.

Subsistence cultivators of the savannah belt also became victims of the rapid expansion of mechanized farming in their region. Lacking the prerequisites, they did not qualify to become scheme operators. Villages were asked to form cooperatives to participate in the scheme, but few of these prospered. The main cause of the peasantry's misfortune was the ensuing shortage of land that led to the abandonment of the traditional cycle of land use, which included shifting cultivation, livestock raising and tree tapping for gum arabic. Of course, land shortage was not caused solely by the invasion of mechanized farming. The high rate of human and animal population growth played a major role. So did the declining productivity of land, itself the result of over-utilization and adverse climatic trends that forced the expansion of cultivation by traditional producers themselves onto marginal land.

What population growth can mean in a local context is illustrated in the case of Northern Darfur Province, whose population this century grew from 130,000 in 1903 to 909,000 in 1973. While the savannah zone is sparsely populated, the distribution of its population is uneven. People congregate where water is available, and there population densities are high. After independence, the state promoted a programme of borehole drilling which enabled clusters of cultivators to appear where no cultivation had been practised earlier, save for the tapping of gum arabic. Mechanized farm schemes were not supposed to be sited near villages, in order not to impinge on local cultivators. However, this rule was often violated, not infrequently by local notables with political influence. In Darfur, between 1960 and 1975, the area under mechanized cultivation expanded from nearly 400,000 hectares to over one million, while yields decreased by one-half (Fouad Ibrahim, 1984, p. 110).

Traditional cultivators were forced to abandon shifting cultivation for lack of land, which meant that fallow periods were shortened or abandoned, except in periods of drought. In turn, this eliminated gum tapping, because new trees had no place to grow, and old ones were cut for firewood. This coincided with a precipitous fall in the price of gum arabic which made it more profitable to use the trees for charcoal (Larson & Bromley, 1991, p. 1294). Continuous cultivation led to a fall in productivity. To compensate for this, the peasants were forced to expand production on marginal lands where productivity was even lower. A vicious circle set in. Unable to meet their full requirements from cultivation and livestock production, many peasant families were compelled to sell their labour. One or more male members of the household found employment in mechanized farms, or joined the army of seasonal workers who journey annually to the riverine region to work in the cotton fields. The loss of male labour diminished further the productive capacity of the peasant household, making it vitally dependent on external sources of income.

Ever since World War II, the Sudanese state has encouraged traditional cultivators to devote part of their land for the production of cash crops. A scheme of regional specialization was adopted, with some regions slated for sesame production and others for groundnuts, gum arabic and sorghum. A growing number of traditional cultivators reduced their production of food staples (*dukhn* and *dura*) to raise cash crops. The cash earned was used to buy various commodities that now became necessities, as well as food in order to supplement their own food production. Thus, on the one hand, staple food production was reduced and, on the other, cash needs were fostered which included the purchase of food.

The mechanized farm scheme did not turn the savannah zone into a granary. In the early years, with good rainfall, the scheme appeared to fulfil its promise, with high yields and high profits for enterprising farmers. In areas where there were no trees to clear and capital investment was modest, it was sometimes possible to recover all the investment in one year. The main crops produced in the early years were *dura* and *dukhn*, the staple food crops, and it was quickly concluded that mechanized farming in the savannah 'has secured the country against famines, which were probable in years of low rainfall, under the old traditional farming systems' (Kamel Mansour, 1972, p. 185). However, disturbing signs could be noticed even then. The scheme rules postulated rotation of crops and land, so that one-third of the land could be left fallow. It was noted that most scheme operators, anxious to make quick profit, did not follow these rules. They exhausted the land through continuous cultivation, then moved elsewhere; they were dubbed 'suitcase farmers'. They also cultivated illegally land that had been classified as unsuitable.

Thorough elimination of vegetation, from trees to weeds, lowered the moisture-retention capacity of the land, and exposed it to water and wind erosion. The latter reached such an advanced stage during the dry period that atmospheric dust reduced visibility in some areas. Severe gully erosion appeared on the clay plains: *kerreb* and *haddam* are two terms used locally to describe this type of land degradation. In the early 1970s, the great Sahelian drought affected the savannah zone, and productivity dropped sharply. In order to maintain output, the area devoted to *dura*, *dukhn*, sesame and groundnuts expanded by more than half, while yields declined steadily. The adverse effects on the people of the region were described by Abdel Ghaffar (1982, p. 47) as follows:

> cultivators are forced to sell their labour cheaply, pastoral nomads are driven out of the best areas of their traditional pasture to places which are not favourable to their herd growth, and agro-pastoralists are being subjected to various socio-economic pressures and forced to abandon one of the two activities and change over to agricultural labour with lower standards of living.

The problem of ecological degradation is not new in Sudan. The first signs were noticed in 1940 in areas already suffering the consequences of pastoralist confinement, over-stocking and over-grazing. A Soil Conservation Committee was formed to study the problem, and its report, published in 1944, cited congestion and over-grazing in most pastoralist areas, including Southern Sudan. The situation in the pastoralist realm has grown steadily worse since then. A Rural Water Corporation was founded in 1956 to supply water for people and animals. A large number of boreholes were dug in the first two decades of independence, turning seasonal pasture into permanent grazing. Land in the vicinity of the boreholes was quickly denuded and degraded. The same happened to routes over which animals trekked.

In the 1970s, the spectre of advancing deserts rose to haunt Northern Sudan. A report by the Desert Encroachment Control and Rehabilitation Programme in the Ministry of Agriculture, published in 1976 by the National Council for Research, put the rate at which the desert was alleged to be advancing at 5–6 kilometres a year. This was bound to conjure up apocalyptic images of the future, as the following excerpt from the report of a scientific mission shows:

> About 60% of the best irrigated agricultural land has been smothered by wind-blown sand, and 40% of the population has left the area (mainly for the Gulf States). The river Nile is gradually moving westwards as a result of erosion. In the long run, there is a possibility that it will change course and flow into a depression with a radically different outlet than its current delta in Egypt. (African Academy of Sciences, 1987, p. 18)

A contrasting point of view holds that people are moving northwards to cultivate the fringes of the desert, rather than the desert moving southwards. According to one estimate, man has pushed millet cultivation in Sudan northwards 200 kilometres past the agronomic limit (Fouad Ibrahim, 1984, p. 110).

Cultivation and pastoralism are practised by the Nilote pastoralists in Southern Sudan. The people there live in permanent villages above flood level and cultivate land around them, using the hoe to plant mostly millet. During the flood, the cattle are moved to high ground where rain-fed pasture becomes available. In the dry season, the herds move to low ground towards the edge of rivers, lakes and swamps that never go dry. Among the Nilotes, the Shilluk, Anuak and Bari have turned entirely to cultivation in recent times, but the memory of their pastoralist past is still vivid in their tradition. The Nilotes supplement their diet by fishing with spears in the receding flood waters.

The agro-pastoralist economy of Southern Sudan underwent little change during the colonial era and afterwards. Several abortive attempts were made during the colonial period to promote cash crop cultivation. The Zande Scheme in Equatoria succeeded, only to be wrecked in the civil war that followed Sudan's independence. In the colonial period, Southern Sudan became an importer of grains to feed the non-producers who congregated around the administrative centres in the region. During the period of self-government in the 1970s, the regional government planned to promote cash crops like tobacco and sugar, but nothing materialized before the second civil war broke out. If there is a positive side to such total lack of development, it could be the preservation of the region's pristine nature and resources. Unfortunately, in addition to heavy loss of life and great human suffering, the war has taken a heavy toll on certain resources. Hardest hit is the livestock of the Nilotes. Cattle are the item most commonly looted by the multitude of armed men who comprise the Sudanese government army, the Southern rebel guerrilla army, several tribal militias and many common outlaws. Heavy damage is done to the environment in areas where people and animals are forced to congregate in search of safety and food.

Increasing pressure on diminishing resources is the source of widespread conflict throughout Sudan, although this is often unheralded because it is fused and confused with larger confrontations. In order to counter the threat posed by the renewed rebellion in the South in the 1980s, the Arab-dominated regimes in Khartoum resorted to arming tribal militias, presumably in order to combat the Sudanese Peoples Liberation Army (SPLA). As is often the case in the Horn, some groups took advantage of the wider conflict to wage their own struggle over diminishing resources with their neighbours. One prominent case involves the Humr Arab pastoralists who share the same district of South Kordofan with the Ngok Dinka Nilotes. Both groups raise cattle and cultivate grain for their own consumption, and have lived in perennial rivalry over pasture, water and migratory routes. Traditional rivalry and raiding escalated into all-out warfare in the wake of the first Southern Sudanese rebellion waged by the *Anya-nya* in the 1960s, as the two groups sided with opposing sides in that conflict. The Humr and Ngok raided each other's cattle camps, taking animals and killing their owners. After a decade of peace in the 1970s, the conflict resumed in the 1980s when Sudan was engulfed once more by civil war. This time the Humr were heavily armed by the state as a militia force. They fell on their Ngok neighbours, burning their villages, killing the men, and taking women and children into slavery, along with all the animals they could seize. This time, it appeared, the Humr were intent on clearing the district of Ngok in order to seize their land (Mohamed Salih, 1990b).

Water

Unlike Egypt, Sudan does not fully use its share of the Nile water under the 1959 agreement. In the 1980s, it used no more than three-quarters of its allocation (Whittington & McClelland, 1992, p. 146). Like most other countries in the Horn, Sudan has ambitious plans to expand the area under irrigation. It embarked on a major project in the late 1970s, which was to become 'one of the causes of the present war and is a factor that fuels its continuation' (Lako, 1992, p. 47). That project is the Jonglei Canal in Southern Sudan.

The Nile's leisurely pace results in high evaporation rates, or, as irrigation experts see it, water loss. Great losses through evaporation are incurred also in dams. For example, 10 billion cubic metres are lost annually in the Aswan Dam. Three times as much water is estimated to evaporate in the Sudd, one of the world's greatest natural swamps. In Southern Sudan, the Nile floods after April, as water from its vast catchment area arrives and is supplemented by tropical rains in the region. It covers a vast area choked by vegetation, where it is estimated half of the floodwater evaporates. This is not an unmitigated loss, because the flood irrigates pastures for the herds of the Nilotes during the dry season, and also provides fish for their diet. Nevertheless, the Nile emerges from the Sudd much diminished. The annual flow of the river further upstream at the Ugandan border is in the order of 30 billion cubic metres, while its flow at Malakal, despite the contribution of the Sobat River, is reduced to some 18 billion cubic metres. The annual loss amounts to 40% of the water held at the Aswan Dam.

Reducing evaporation loss is one way of increasing water supply, and the Sudd has long been considered the place where this could be done by digging a canal to bypass the swamps. Such a canal would lower the flood level of the river, reducing the spillover. The idea was first conceived by the British in 1904, and since then many alternative designs have been drawn. The 1959 agreement between Sudan and Egypt provided for several projects in the Upper Nile and Bahr el Ghazal, which would increase the Nile flow by 10 to 15 billion cubic metres annually, making it possible for Sudan and Egypt to irrigate an additional 1.2 to 1.6 million hectares each. The Jonglei scheme was estimated to save 3.8 billion cubic metres of water, to be shared equally by Sudan and Egypt. The first civil war in Sudan delayed implementation of these projects but, soon after the 1972 Addis Ababa Agreement was signed ending that conflict temporarily, the Permanent Joint Technical Committee (formed by Sudan and Egypt in 1959) produced the final plan for the Jonglei Canal. (Jonglei is a Dinka village 80 kilometres north of Bor, where the canal began its 280 kilometre course to Malakal.) A unique, giant earthmoving machine was imported, and work began in 1978.

As one observer noted, whereas a great deal of research went into the hydrological and engineering aspects of the Jonglei Canal in advance, little thought was given to the potential ecological impact of the scheme and its socio-economic effects on the people of the region. Until, that is, the project became the object of widespread controversy (Lako, 1992, p. 45). It was estimated the canal would reduce the flooded area by one-fifth, and this was perceived as a gain of cultivable land (1.5 million hectares) that could be used to modernize agriculture in the South. A modern agricultural sector would, among other benefits, provide local employment for an area that exports a large part of its labour force. Improved transportation through a most difficult area was another perceived benefit for the region, while the canal itself was presented as a new fishery source. Such incidental gains were sufficient to gain the support of most politicians in the then self-governing Southern Region. Students in the South, however, perceived this as another high-handed imposition from the North, and demonstrated against the scheme in 1974. The Southern regional government was unimpressed. Its head, Abel Alier, told the regional assembly: 'If we have to drive our people to paradise with sticks, we will do so for their good and the good of those who come after us' (Lako, 1985, p. 29).

Subsequent independent studies raised many questions about the potential impact of the scheme on the region and its people. The rationale of the scheme itself was challenged by those who pointed out that at least part of the envisaged water gain would be cancelled by increased evaporation over an expanded area at Aswan. It was argued also that the swamps play a regulating role in the flow of the Nile, and that reducing them could increase the severity of flooding. The perceived benefits for the region were challenged by others who argued there is no shortage of cultivable land in the South, that intensive cultivation in the modern sector would provide little local employment, and the loss of pastures would be a serious blow to the pastoralist economy. Moreover, pastoralist mobility and access to water would be impeded by modern farming schemes, not to mention the presence of the canal. Even fishing resources could be negatively affected, it was alleged, since certain species breed in the shallow floodplains, and reducing these floodplains would diminish the fish population.

The most serious questions were raised by environmentalists who objected to the launching of the project without proper study of its potential ecological impact. They raised the fear that the destruction of the wetland ecosystem could trigger major environmental changes in the region, affecting weather patterns and possibly reducing precipitation in the catchment area, interrupting the food chain of the region's fauna, affecting its flora predictably by reducing groundwater recharge and in unforeseen ways by eliminating the swamps' filter role. The Nile floodplain is the habitat of a wide variety of wildlife. Nearly half a million wild horses living there are one of the world's largest remaining population of wild large mammals. Jonglei is a refuge for a huge population of Nile *lechwe* (small antelope), the shoebill stork and probably the largest population of water birds in Africa. Reduction of the floodplain would threaten all these animals, and the obstruction of their migratory movements posed by the canal itself would be an additional hazard.

Southern Sudanese have been taught by history to mistrust the motives of Northern Sudanese and Egyptians. They suspect the Arabs perceive the South as a land reserve for their surplus peasantry, and it was rumours that Egyptians were to settle along the canal that sparked the student riots in Juba in November 1974. Not surprisingly, the canal became one of the first targets of the Sudanese Peoples Liberation Army (SPLA) when the second round of the civil war began in the early 1980s. For the Southern rebel army the Sudd has strategic significance, because it forms a defensive barrier that greatly complicates the transportation problems faced by the invading Northern Sudanese armies, and the canal was to have a road built alongside it. The canal was dug more than half-way when work on it halted in 1983.

In areas away from the Nile in Northern Sudan, water supply for domestic purposes is a chronic problem. The World Health Organization estimate of the minimum daily water requirement for a normal life is 4 gallons: average per capita water consumption in Sudan is about 1.5 gallons. The need for water became so urgent that the state was compelled to embark on an 'anti-thirst' campaign during 1966–69, when thousands of water points were opened in Northern Sudan. Human and animal concentration around these points resulted in land degradation and loss of cover. Sudan's underground resources are believed to lie mainly under the Nubian sandstone and in aquifers in the Gezira formation and the Umm Ruwaba area. Nearly half the country has a basement complex with very low ground water capability.

Minerals and Energy

Sudan's vast territory has yet to be fully surveyed for its mineral wealth. Gold and silver are known to have been produced in the past, but there is no significant production of these precious metals now. There is modest production and export of iron and manganese, as well as salt. Copper, mica and gypsum deposits are known to exist. Sudan has an agreement (1974) with Saudi Arabia to explore and exploit jointly the mineral resources of the Red Sea. The most valuable mineral resource discovered to date is petroleum; that discovery was to become yet another bone of contention between North and South.

Early on, oil exploration in Sudan was hampered by the Petroleum Resources Development Act (1958), which demanded as much as 70% of profits for the state. Until the early 1970s, exploration efforts in Sudan were concentrated in the Red Sea area, while the Southern Region was engulfed in civil war. An Italian company, Agip Mineraria, obtained an exploration licence in 1959. It found natural gas at Debarwa, but not of commercial quantity. In 1974, two years after the first round of the civil war ended, Chevron Oil Company obtained an exploratory licence over an area covering half a million square kilometres in southern Darfur, southern Kordofan, Blue Nile and Upper Nile Provinces. This last named province is in the Southern Region. Information obtained by the United States through remote sensing indicated large deposits of petroleum in the south and southwest. Chevron first struck oil in 1979, at Abu Gabra near Muglad in southern Darfur. The following year, another and more important discovery was made inside Upper Nile Province near Bentiu. More discoveries were made in Bentiu and in the Machar Marshes within the Southern Region. In 1983, Chevron estimated the oil reserves of the Southern fields at nearly one billion barrels, of which 240 million barrels could be recovered.

The discovery of petroleum was of immense importance to a country whose commercial energy consumption is one of the world's lowest, and whose energy needs are met by biomass fuel (up to 80%). Imported oil provides most of the rest, for which Sudan expends more than half of its foreign exchange earnings each year. Hydroelectricity provides only 1% of energy consumption. Wood and charcoal supply the energy for households and some industries like bakeries and brickmaking. Household energy consumption (cooking) is considered the main cause of deforestation in Sudan. Deforestation in the 1980s was estimated to have stripped 104,000 square kilometres, while reforestation restored 11,000 square kilometres (Pearce & Turner, 1990). Khartoum is supplied with woodfuel from areas some 500 kilometres distant.

The bulk of the oil reserves are located within the Southern Region and, at the very least, the Southern Regional government was entitled to collect taxes on oil production. Apparently wishing to preempt Southern demands for a share of oil wealth, the Nimeiry regime in Khartoum tried to disguise the location of the oil fields by naming them Unity One, Two etc., and referring to their location as 450 kilometres south of Khartoum. There was even an attempt to redraw the provincial boundaries in order to create a new Unity Province to incorporate the oil-producing area.

The regional government in the South was not involved in any of the negotiations with Chevron, nor consulted in the decision to establish an oil refinery at Kosti, 550 kilometres to the north. This decision provoked widespread resentment and student demonstrations in the South, whose regional government wanted the refinery to be based at Bentiu. The regime in Khartoum then announced a change of plans and the intention to build a 1,425 kilometre pipeline to take the crude oil to a new terminal on the Red Sea between Port Sudan and Swakin, from where it was to be sent abroad for refining. In justification of this plan, it was said that refining capacity worldwide was 40% idle, and refining prices were so low that it was cheaper to have the oil from Bentiu refined abroad than to build a new refinery at Kosti (*Sudanow*, March 1983). The oil wells at Bentiu were supposed to go into production in 1985. A year earlier, the Sudan Peoples Liberation Army forced Chevron to abandon the area. Ten years later, the oil fields remained abandoned.

3.4 Somalia

The Somali are the largest group of traditional pastoralists in the Horn of Africa, and one of the largest in the world. Nomadic pastoralism without permanent habitation or cultivation has been the vocation of the Somali throughout their history, and the pastoralist ethos pervades their culture. The nomads built no states of their own, nor recognized any foreign masters. Political authority among them was diffuse in the extreme, and their society was described as 'democratic to the point of anarchy' (Lewis, 1980, p. 10). The trading towns on the Somali coast of the Red Sea and Indian Ocean, and the sedentary Saab cultivators in the Juba–Shebeli valley, differed from the pastoralists in their economic pursuits and sociopolitical organization, and lived in a state of constant tension with the nomads of the interior.

Little allowance was made for the distinctive nature of Somali society by its rulers during the colonial and post-colonial period. The colonial economy throughout the Horn was founded on the production of cash crops for export. Capital investment and technological innovation were concentrated into relatively few areas where intensive cultivation was possible, which were also areas of high population density and plentiful supply of cheap labour. Under Italian colonial rule, the Juba–Shebeli valley in the south became the focus of economic development. Capital was invested in the production of bananas, cotton, sugar and other crops grown on irrigated land. About 40,000 hectares were occupied, with local cultivators providing the labour force. State investment in infrastructure, direct subsidies and a market monopoly in Italy where the price of the fruit was twice the world price, made banana production a success and the mainstay of the colonial economy.

In the 1950s, banana exports accounted for 60% of total export value. In British Somaliland in the north, where the arid terrain offered no similar possibilities for commercial cultivation, there was no development of any sort, and pastoralism continued to reign supreme.

Commercial cultivation remained the focus of economic development efforts of the Somali state, which united the two former colonies in the wake of independence in 1960. Capital investment was concentrated in expanding and improving the production of bananas, the crop that provided the state with its main source of revenue and foreign exchange. Investment came from private sources, mainly Italian, but a small number of Somali also gained a foothold in this sector. Banana production increased by more than 50% in the first five years of independence. State investment also went into sugarcane production on irrigated land in the Shebeli region. In all, 20% of development expenditure went to commercial agriculture.

By contrast, only 3% was earmarked for animal husbandry, mostly to promote marketing. The opening of a major market for live animals, mainly sheep and goats, in the Arabian Peninsula in the 1960s stimulated the state's interest in livestock marketing. Water supplies expanded through a programme of borehole drilling and artificial pond construction, stock routes and holding grounds were improved, and veterinary services were expanded. Exports of animals increased rapidly. Together with animal products, the livestock sector took the lead in the export market that had till now been dominated by bananas, providing the state with an additional source of revenue and foreign exchange.

The inherent pastoralist tendency to expand the herd was reinforced by the pull of the market, and facilitated by the provision of water supplies and veterinary services. The result was rapid growth of animal population, over-grazing, especially near water points, and land degradation. An FAO survey in 1967 estimated the national herd had increased almost three times since the late 1950s, and an ILCA mission in 1971 estimated the number of animals had increased 40 to 60% in the preceding five years. Livestock suffered disastrous losses in the drought of the mid-1970s, yet by 1981 the size of the national herd was estimated at more than 41 million (Somalia, 1988). The northern region had the bulk of flock animals; cattle were concentrated in the centre and south, while camels were distributed throughout the country.

Pastoralist production in Somalia rests on a precarious base. More than 10% of the total land area is non-productive even for livestock, and over 40% offers only limited possibilities for grazing. The entire territory of Somalia is highly vulnerable to drought. Records and popular memory recall 14 major incidents of drought between 1911 and 1973; that is, one every four to five years (Swift, 1977). Limited and highly erratic precipitation, short periods of pasture growth, lack of permanent natural water sources, long distances between man-made wells, vegetation of limited nutritional value, and other handicaps of nature require constant movement, wide dispersal, as well as exceptional skill in man and adaptive capacity in animals to maintain a precarious balance between humans, animals and natural resources. In the past few decades, human and animal proliferation upset this balance, starting a ruinous process which was further accelerated by adverse climatic trends.

Animal proliferation made over-grazing unavoidable. Traditional practices, such as reserving certain areas for grazing during only part of the season, could not be maintained. The presence of wells became an inducement to stay long periods in the same area, and herders tended to congregate in semi-permanent camps and villages. Continuous grazing leads to the destruction of preferred species of vegetation, because the animals devour not only the leaves but the stems and inflorescences as well, before turning to other species in the order of preference. This process leads to the disappearance of most sought-after species and the proliferation of others with little or no nutritional value. In the worst case, congestion and over-grazing lead to land denudation and degradation.

This malignant process was noticed already during the colonial period. A 1947 official report in British Somaliland warned that 'the soil and vegetation are on the brink of irretrievable ruin' on account of animal congestion and over-grazing (Geshekter, 1982, p. 23). The colonial government established reserve areas where grazing was restricted and tree felling was prohibited. In the 1960s and 1970s, borehole drilling by the state and privately constructed concrete water-tanks accelerated the process of land degradation, until the state was obliged to prohibit the drilling of wells. The drought of the early 1970s destroyed a large portion of the herds, especially in northern Somalia, and it also further weakened the fragile environment. A major range rehabilitation project for the northern rangeland was launched by the World Bank in 1976, but made little progress. The war with Ethiopia in the later part of that decade brought hundreds of thousands of refugees from the Ogaden with their animals into Somalia, mostly into the north. The impact on the environment was devastating.

The shrinking resource base in northern Somalia had already provoked a major conflict between the Ishaq, the dominant clan family in that region, and the Ogaden clan that lives across the border in Ethiopia. These two Somali clans had been feuding over the rich Haud pastures inside Ethiopia for several decades. Although it lacks permanent water supplies, the Haud offers rich seasonal grazing dependent on rainfall. In the past, the lack of permanent water sources ensured the area was vacated by herders during the dry season, thus escaping over-grazing and preserving its full potential. It was visited seasonally by the Ogaden clansmen in Ethiopia and the Ishaq from northern Somalia, as well as other Somali clans. During the colonial period, the British authorities in Somaliland secured an agreement with Ethiopia on behalf of the Ishaq, designating an area of 25,000 square kilometres into which the latter could move their herds seasonally.

During World War II, the British took control of the Haud and kept it until 1954, when it was returned to Ethiopia. During that period, the Ishaq took advantage of British protection to make inroads into the Haud, prolonging their stay there and, when conditions permitted, remaining there throughout the year. The Ogaden found themselves pushed westwards away from the

Haud and reacted violently, setting off a conflict between these two clans that escalated greatly when it became woven into larger confrontations. In the mid-1970s, the Ogaden and the Haud were infiltrated by the Western Somalia Liberation Front (WSLF), whose goal was the liberation of the Ogaden from Ethiopian rule and its incorporation into the Somali Republic. The Ogaden clan provided the WSLF with its fighters, and they took the opportunity to wreak vengeance on the Ishaq who ventured into the Haud. The Ishaq had their turn a decade later, when the Somali National Movement (SNM) was formed to struggle against the Siad Barre regime in Mogadisho. The SNM was an Ishaq organization, and when the Ethiopians allowed it to operate inside the Ogaden, it made certain the Haud became a closed area for the Ogaden clan. Thus, a pastoralist feud over resources was fought under the umbrella of larger conflicts within and between states.

Over-grazing and tree cutting gave rise to a different problem in southern Somalia: the menace of advancing sands from destabilized dunes in the coastal region of the south, where the country's best agricultural land is located. Hundreds of thousands of square kilometres of dunes are on the move, with the dried-out sand being lifted by heavy winds to form dunes as far as 30–40 kilometres inland. In the past, the dunes were covered with vegetation of herbs and bushes sufficient to resist the wind and to hold down the sand. However, these were destroyed by over-grazing, and their place was taken by flora that is useful neither as fodder nor for holding the sand. A contributing factor was the presence of a string of more than 100 wells along the coast between Mogadisho and Kisimayo, which marked a migration route for herds across the fragile dunes. In the early 1970s, when the sands were threatening to block the road south of the capital, the Siad Barre regime launched a dune stabilization campaign. The campaign did not last long, nor did it accomplish much; and after two decades of inaction, the menace of advancing sands was greater than ever.

Nearly one-third of the total land area of Somalia offers marginal opportunities for cultivation, as well as good grazing. About one million hectares (12% of the potentially cultivable land) is under rainfed cultivation in these marginal areas, having doubled in size since 1970. In the south, this is concentrated on the elevated plateau (over 500 metres) north of Baidoa, where mean rainfall is about 500 mm but subject to great variation. Sorghum is the main crop planted in this region, supplemented with maize, beans, cowpeas and sesame. A type of bush fallow is practised when the land is exhausted after several years of continuous cultivation. Simple tools are used to till the land: hoes, *kawawas* (wooden land-levellers and ridge-makers) and sharpened sticks. Animal traction is used only in the northwestern corner of northern Somalia, especially in the Borama and Gabileh districts, where former pastoralists have taken up cultivation following the example of their Oromo neighbours. Elevations there range from 600 to 1,500 metres, and precipitation varies between 500 and 800 mm. Sorghum and maize were the main crops planted earlier, but these are now overtaken by *khat*, a plant with a mildly stimulating effect which is in great demand throughout the Horn and across the Red Sea (Abdi Ismail Samatar, 1989, ch. 5).

Irrigated cultivation is practised in the Juba–Shebeli valley, where controlled flooding in the alluvial riverine plains offers the best agricultural potential in the entire country. The total area potentially available for irrigation is estimated at 250,000 hectares. Of this, nearly one-fourth has been used to grow sugar, bananas, rice, citrus fruits, maize and sorghum. The Shebeli River is intensively used; because so much water is drawn from it for irrigation, it dries out before reaching the sea. In the Juba region, only 20,000 hectares out of an estimated potential 200,000 hectares have been used. The unfinished Bardhere Dam was designed to fully utilize the water of this river.

Salinization is a serious environmental problem related to irrigation in the Shebeli river region, because the river has a high saline content even during high flow. Irrigation causes salinity to increase by raising water tables, thereby bringing dissolved salts into contact with plant root systems and eventually the land surface, where evaporation fails to remove salt which accumulates as a result. Salinization is particularly severe in poorly drained land developed during the colonial period. Much of this land now is under-utilized or not utilized at all. Productivity in the Jowhar sugar factory fields was seriously affected by salinization, and thousands of hectares of land had to be abandoned. Sugar yields per acre fell from 40 tons of cane in the 1960s to 22 in the 1980s. Generally, yields in land under irrigation are far below international standards.

Water

The fact that water is a scarce resource throughout Somalia is attested to by the number of official bodies that have been concerned with it. These included a Water Development Agency in the Ministry of Mineral and Water Resources, a National Water Committee, a National Technical Committee, and numerous local agencies. With two-thirds of the population lacking access to clean water, improving urban and rural water supplies was their main task. Given the scarcity of surface water, underground water was the focus of their efforts. Until the 1980s, only three towns – Mogadisho, Kisimayo and Hargeisa – had water-supply systems. Mogadisho gets its water from a large number of deep wells, Hargeisa from six wells, while Kisimayo gets its water from the Juba River. Four more towns in the south, and Berbera in the north, acquired water-supply systems during the 1980s. There is no urban sewage system in the country. A beginning was made in 1983 to construct one in Mogadisho for the use of 10,000 people, when the city's population was estimated at 700,000.

In rural areas, people collect surface water in tanks, basins and weirs, and resort to shallow wells when surface water is not available. Wells, a perennial feature in all rural settlements, are also used for irrigated cultivation. For a time, the state promoted borehole drilling throughout rural Somalia, until it was forced to abandon this practice when the environmental consequences became obvious. Privately owned cement tanks also proliferated in the bush, where water is purchased by the pastoralists, and private well drilling expanded later without being subject to any controls.

According to an official assessment, 'the development of rainfed agriculture is approaching the limits of its potential because of the lack of alternative crops suitable for production under local conditions. The major development and growth in the sector will be in irrigated production, since demand for the crops produced is large, land suitable for irrigation is still available, and improvements in technology and management could increase output substantially' (Somalia, 1987, p. 165). This assessment was based on the calculation that production of sorghum and maize at that time exceeded local demand.

The expansion of irrigation was to be accomplished through better water management in existing schemes, and the construction of new schemes. Efficiency of existing schemes is rated very low (20–30%). It has been calculated that efficiency in the use of the water from the Shebeli River, whose estimated annual flow is over 2 million cubic metres, could improve by 45% to meet current demand. The World Bank Agricultural Sector Review recommended that all expansion along that river cease and efforts be concentrated on improving efficiency (World Bank, 1985). Hopes for expansion of irrigated cultivation then focused on the Juba River, whose estimated annual flow is over 6 million cubic metres. This flow has extreme seasonal fluctuations, ranging from a peak of 1,500 cubic metres per second to a trickle. The need to regulate the Juba was recognized as early as the 1920s by the Italian colonial administration. Plans for a giant dam on the upper Juba, 35 kilometres upstream from Barhere, were made in the second half of the 1980s, and provision was made in the *Five Year National Development Plan, 1987–1991*. The dam was to provide water regulation for the Juba, to make it possible to triple the area under irrigated cultivation in the Juba region, and to provide hydroelectric power generation. No study was made of the potential impact of the dam downstream or along the coast where the river ends. Lack of funds led to the abandonment of this project.

Minerals and Energy

Mineral exploration has been going on for half a century in Somalia, with meagre results. The quarrying of industrial minerals (sand, gravel, limestone, gypsum, clays) for use in construction constitutes the only mining activity in the country. The extraction of limestone and of sand close to the shoreline near the main coastal towns is creating an incipient environmental problem. A tin mine operated during 1972–79 at Bosaso, producing only a few tons of processed ore. The existence of iron has been established in El Bur, uranium in Galgaduud, and high quality (piezo-electric) quartz in the northwest. Several occurrences of oil and gas have been confirmed, but their commercial potential has not been proved. The military regime that came to power in 1969 imposed stringent conditions on exploration by foreign companies. The *Three Year Development Plan, 1979–1981* confidently asserted 'the greater part of the work in this sector will in future be accomplished without having to rely on foreign assistance' (Somalia, 1979, p. 132). Several years later, the *Five Year National Development Plan, 1987–1991* recorded zero progress, adding that 'several of the projects to investigate and develop mineral resources envisaged in the past plan period failed to attract the necessary financial support' (Somalia, 1987, p. 205).

Somalia's energy needs are met mainly (82%) by woodfuels. Estimates of Somalia's forest cover vary wildly, depending upon the source and the criteria used. Official estimates refer to 52,000 hectares of 'dense' forest, and 5.7 million hectares of 'low density woodland' (Somalia, 1987, ch. 7). Besides providing Somalia's energy and construction needs, woodland produces the fabled frankincense and myrrh resins, as well as gum arabic. Urban demand for charcoal and building material has increased steadily, following demographic trends which are reversing the traditional Somali nomadic way of life. A stark example is northern Somalia, an area bereft of inland towns in the pre-colonial era, whose capital, Hargeisa, reached a population nearing 400,000 in the 1980s. (It lost much of it after the destruction it suffered in the civil war at the end of that decade.) Wooded cover throughout Somalia shrunk correspondingly with this trend, with relative scarcity reflected in rising wood and charcoal prices. Petroleum imports, though contributing only a small portion of energy consumption, consumed approximately 40% of export revenue. Petrol has been used to generate electricity for the urban centres and in industry.

3.5 Kenya

Kenya has the most advanced commercial agriculture in the Horn of Africa. It is Africa's leading tea exporter and the third largest tea-exporting country in the world, accounting for about 10% of the world market. Kenya is also a leading exporter of coffee, and produces nearly 70% of the world's pyrethrum. It also exports wheat, sisal, sugar, beef and dairy products, fruits and vegetables. Commercial agriculture is the most dynamic sector of the national economy, accounting for about two-thirds of the country's total foreign exchange earnings. Kenya's hopes for further development are concentrated on this sector, which is the focus of state planning and investment as well as private initiative.

The foundations of commercial agriculture in Kenya were laid early during the colonial period, when Europeans, whose number ultimately reached about 4,000, settled in the highlands. A huge area of more than 3 million hectares of the best land was set aside for them, pushing out the African peasantry and laying the basis for the land problem that has haunted Kenya ever since. The settlers carved out very large farms, and set about producing commercial crops with the labour of dispossessed African peasants. The Europeans secured a production monopoly by preventing, through various means, African production of commercial crops. Africans were allowed to produce a little cotton and the wattle bark used for tanning leather that is one of Kenya's staple exports. Because they also produced beef, the settlers imposed a quarantine that prevented the marketing of animals from Kenya's northern lowlands in the White Highlands.

The Mau Mau rebellion in the early 1950s was the direct result of the severe land shortage experienced by Africans, particularly the Kikuyu, due to the expropriation of their lands by the settlers, and led to a radical change of land policy on the part of the colonial administration. It now sought to promote the emergence of an African middle farmer class through various methods, including land consolidation, private ownership registration, credit provision, extension services, and allowing Africans to compete with European settlers in the production of commercial crops. Thus were laid the foundations for the development of indigenous commercial agriculture. This development advanced rapidly after independence in 1963. Most of the White Highlands were Africanized and to a large extent sub-divided, expanding the base of the indigenous commercial farmer class.

Opening the commercial cultivation sector to Africans had impressive results. Overall agricultural production growth rates as high as 6.2% were registered during the first decade of independence, outpacing a 3.6 rate of population growth. Production increases in commercial crops, especially tea, were even higher, and per capita income rose by more than 3% annually. The expansion of this sector continued in the decades that followed. New land was continuously brought under cultivation. Tea, cotton, sugar, maize and wheat production was greatly expanded, and horticulture emerged as a major export producer. Kenya's economic performance was high enough above sub-Saharan Africa standards to be considered a minor economic miracle.

Such agricultural expansion was bound to come up against the limits of Kenya's natural endowment. Kenya's entire territory covers 582,646 square kilometres, of which 11,230 square kilometres is covered by water. About 70% of the land is classified arid, suitable only for livestock grazing of varying degrees of intensity. Of the remainder, less than 10% is classified as land of high potential in the sense of having adequate rainfall and good soils, and the rest is of medium potential. Most agricultural activity is concentrated in the high and medium potential zones, which account for about 18% of the total land area and are inhabited by two-thirds of the country's population. Given the high rate of population growth, population density on arable land is reaching the saturation point, having risen from 123 per square kilometre in 1979 to 209 in 1993 (Kenya, 1989, p. 171).

Inevitably, population pressure in the medium and high potential areas led to the spread of cultivation into the semi-arid and arid areas, where the ecology is fragile and crop production risky. State planning has focused on these vast regions which contain about a quarter of Kenya's population. The Arid and Semi-Arid Lands (ASAL) Programme was launched in 1979 to promote dry-land farming, using drought-resistant crops, small-scale irrigation projects and ranching in areas used until now by mobile pastoralists. A Ministry for Reclamation and Development of Arid and Semi-Arid and

Wastelands was established. Irrigation offers another possibility for agricultural expansion. Kenya has an estimated 500,000 hectares of irrigation potential, and another 300,000 hectares amenable to drainage and valley-bottom reclamation.

Since there are no other possibilities for significant cultivation expansion in Kenya, the runaway rate of population growth is shaping up as a major problem for the foreseeable future. Agriculture will continue to bear the burden of feeding a population which expanded at the rate of one million per year during the 1988–93 plan period, and to provide employment for at least three-quarters of the labour force. Arable land per capita fell from 0.2 hectares in 1965 to 0.1 hectares in 1987, and if present trends persist it will decline to 0.07 hectares by the end of the century. Understandably, land has great social and economic value, and is the subject of intense competition waged at various venues and levels: 'Everywhere there is competition for limited land and water resources' (ODA, 1994, p. 1) 'Land issues are central to Kenya's social system and will be for years to come' (Miller, 1984, p. 86). Land is the subject of massive litigation between individuals, and of conflicting claims between villages, communities and ethnic groups which retain bitter memories of territories won and lost in times past. For example, the Kalenjin attacks against the Kikuyu in the Rift Valley in 1993 were partly spurred by the resentment of the former against the intrusion of the latter into the region when the European settler estates were taken over. The holdings of the European settlers in Kenya were not equally divided. Large estates passed into the hands of Kenya's ruling class, and these are not always used to full capacity. On the other hand, an increasing number of Kenyans are joining the ranks of the landless. This imbalance makes land an explosive political issue.

Given the limits imposed by nature on horizontal expansion of cultivation, Kenya's efforts have been concentrated on boosting productivity through greater use of inputs. 'Since with rapidly increasing population good agricultural land is becoming scarce, the challenge faced by the agricultural sector . . . lies in effective promotion of increased and widespread use of improved seed varieties, fertilizers and pest and disease control chemicals' (Kenya, 1989, p. 119). Hybrid maize breeding in Kenya began in the 1950s, and several varieties have been developed to suit climatic and soil conditions, gradually displacing the indigenous varieties that provide the staple food crop. 'Hybrid maize has become so popular in Kenya that most farmers in the High Potential and Medium Potential ecozones will not accept anything else' (Ochieng, 1985, p. 27). Improved seed production is also planned for barley, beans, potatoes, horticultural crops and grasses. Fertilizer is the dominant farm input used; with a consumption rate of 51 kilograms per hectare of arable land and permanent crops, Kenya outstripped Sudan (4 kg), Ethiopia (6 kg) and Somalia (2 kg) (FAO, 1991b, p. 212). The 1989–1993 development plan included various incentives designed to increase fertilizer use, particularly among small farmers.

The heavy demands placed on it are taking a toll on the land. 'In Kenya, land degradation through soil erosion is a pressing agricultural problem presenting a major threat to all facets of land productivity . . . Land in some parts of Kenya has already been abused to such a degree as to limit severely the options for future productive use' (Kilewe & Thomas, 1992, p. 1). The danger of soil erosion is great where cultivation has crept up the slopes of mountains, as with Mt Kenya, also where cultivation has expanded on streambanks and along stock-tracks and footpaths. Evidence is found not only on the surface of the land, but also on the coast where river deltas have expanded and formerly sandy beaches have turned into mud flats, menacing marine life in estuaries and lagoons, and endangering the coral reef by turbidity and loss of light. For example, sediment discharge from the Sabaki River has affected beaches and coral reef near Malindi, where beach accretion was measured at up to 500 metres (UNEP, 1982, p. 9).

Fertilizers, pesticides and other agro-chemical inputs pose also a threat to people and the environment. One study (Christianson, 1991) found that many pesticides banned in industrialized countries due to their ill-effects on humans and the environment were widely used in Kenya, where farmers had little knowledge concerning their harmful potential or proper handling. It has been officially acknowledged that fertilizer use 'has had severe pollution effects in some cases', particularly in the coffee-growing areas (Kenya, 1989, p. 175). Fertilizer wash-out, assisted by soil erosion, enters small creeks and rivers which collect all surface run-off, and ultimately reaches the sea. There, it produces a concentration of nitrogenous and phosphorous compounds which stimulate eutrophication in estuary and lagoon waters.

A very large portion of Kenya's arid territory is devoted to livestock production practised by mobile pastoralists, who own approximately half the nation's animals. The Rift Valley, the savannah plains of the highlands and the northern lowlands are pastoralist domain, although cultivation has been intruding there since the colonial period. The Masai displacement from the northern Rift Valley and the Chamus and Tugen displacement in the Baringo district to make room for European settlement, are well-known cases of massive encroachment into the pastoralist domain. Out-migration from the overcrowded areas of the highlands is constantly bringing cultivators into marginal areas formerly used for grazing. The northern lowlands, which represent 40% of Kenya's territory, are least affected by this trend, although numerous small-scale irrigated cultivation schemes have been founded in recent years to accommodate landless peasants and impoverished pastoralists.

Tourism is the leading earner of foreign exchange in Kenya, and wildlife resources are the major attraction. In order to protect this source of revenue, Kenya has taken the lead in Africa in the field of wildlife protection. National parks, game reserves and sanctuaries occupy 7.5% of the total land area. Most of this land is found in the semi-arid region where pastoralists used to roam. If they are allowed to use this land at all, pastoralists are subject to many restrictions. Their animals have to compete with the wildlife that congregates there, and run a higher risk of disease infection because of the wide range of livestock diseases that originate in wildlife species.

Land degradation became evident during the colonial period. Grazing control schemes were introduced in the late 1940s to encourage stock limitation and rotation of grazing, and to keep stock away from permanent water sources around which the land was turning to desert. These schemes provided water, dipping and vaccination services, as a result of which livestock numbers in the schemes increased rapidly. Instead of selling the surplus animals, as they were required to do by scheme rules, most pastoralists preferred to quit the schemes; alternatively, they sent the extra animals to be looked after by kinsmen elsewhere. The overall results were negative.

Animal congestion and over-grazing are the familiar causes, with periodic drought adding momentum to this process. Loss of mobility is a critical factor. For obvious reasons, pastoralists tend to congregate around aid distribution centres, also near villages and government stations because of increasing violence and animal raiding in the bush. The prevailing poverty is forcing men to seek menial jobs in settled areas, creating a labour shortage in the pastoralist milieu which results in restricted mobility for herds left in the care of women, children and the elderly. Pastoralist practices, like burning the range and tree cutting for enclosure construction, may also have contributed to land degradation.

As in all other countries of the Horn, development planners in Kenya do not believe the traditional pastoralist mode of production has the potential for development within its own framework. The exploitation of vast areas by small groups of herders who contribute only minimally to the market is considered wasteful. Moreover, pastoralist practices, particularly herd expansion, are regarded as destructive of the environment. Several attempts have been made to promote commercialization of livestock production in the arid region in order to increase the supply of cheap meat for the urban sector and the export market. In the 1970s, the Group Ranching Scheme was introduced to promote this aim. Groups of pastoralist families were given title to an area of land and were provided with credit, marketing facilities, water supply and veterinary services. However, the pastoralists refused to stay put, or to limit the size of their herds, and the result has been described as 'a howling failure' (Graham, 1989, p. 185). One consequence was the beginning of land privatization, at least among the Masai, when group schemes dissolved and ranches were subdivided.

Ultimately, the planners came to the conclusion that 'the possibilities for increasing livestock production lie mainly in intensive feeding and zero-grazing' (Kenya, 1989, p. 121). Commercial ranching by private enterprise was the approach espoused by the 1989–1993 development plan. The so-called 'feeding lots' and 'fattening ranches' buy young animals from pastoralists and fatten them for the market. In addition to increasing market supply, it is believed this will limit over-stocking by the pastoralists and will benefit the environment.

Water

Kenya has 11,230 square kilometres of open water sources in lakes, rivers and dams. The main rivers radiate from the central dome formed by the highlands, and lesser bodies of water descend from the southern foothills of the Ethiopian highlands. Various types of basins and depressions collect water to form lakes. Overall, it is considered the country has enough water to satisfy human and animal needs, agricultural and industrial requirements, as well as hydroelectric power generation. However, these needs are not satisfied at present. Only 35% of the population has access to clean water supplies at a reasonable distance. This includes 75% of the urban population and 26% of the rural dwellers (Kenya, 1989, p. 87).

Water resources are unevenly distributed, with relative abundance in the highlands and the lake regions, and scarcity in the lowlands. Kenya has not yet developed a long-term plan for the use of irrigation in agriculture, though it is estimated it has an irrigation potential of half a million hectares. At present, only 36,000 hectares are under irrigation. Under the Arid and Semi-Arid Lands Programme that began in 1979, irrigation has been introduced in the lowlands for the purpose of settling the surplus peasantry from the highlands and pastoralists who have lost their animals.

Large-scale irrigation schemes were deemed expensive to build and to maintain, and preference was given to small-scale schemes, self-managed by smallholders, with technical and advisory support from the state. Numerous schemes were founded along the Ewaso Nyiro and Tana Rivers in the northern lowlands, and along the Turkwel and Kerio Rivers in the Turkana district. The performance of the small irrigation schemes was not very encouraging, and the focus shifted towards 'water harvesting'. This involves a variety of techniques of water conservation, including the collection of run-off from roofs, rock catchments, land surfaces and seasonal streams, diversion of flash floods etc. (Hogg, 1988).

Minerals and Energy

No major mineral finds have occurred as yet in Kenya, although foreign companies have been prospecting for copper, nickel, chrome and gold. Numerous mineral occurrences have been identified. Those of economic significance include soda ash, fluorspar and salt. Limestone is mined to produce cement, and other construction materials are quarried.

In common with the other countries of the Horn, Kenya has depended mainly on woodfuel (70%) to meet its energy needs. Three-quarters of the charcoal used in urban areas was derived from rangeland vegetation and Kenya has also exported charcoal to Saudi Arabia and the Gulf states, fetching high prices. True forest covers 3.4% of the total land area, and nearly all of it is found on the high-potential agricultural land on the highlands, where it has been gradually eaten up not only by creeping cultivation, but also by energy demand. Some 28% of annual woodfuel production has come from gazetted forests, 47% from agro-forestry and 25% from rangelands. Forest wood has also been consumed by the construction industry, which required 600,000 cubic metres of logs annually. The wood panel industry required another 80,000 cubic metres, with additional demand coming from a pulp paper mill at Webuye. The impact on the environment has been disastrous. Kenya had by far the highest rate (1.7 per annum) of deforestation in the Horn in the 1980s, compared with Sudan (0.2) and Ethiopia (0.3) (Pearce & Turner, 1990). Deforestation was estimated to have stripped 39,000 square kilometres in the 1980s: 'The consequences of widespread deforestation and the laying bare of the rangelands through harvesting of fuelwood are extremely serious' (Kenya, 1989, p. 179). The rate of decrease in supply in this area has been estimated at 1% per annum. Supply shortfall is expected to reach 30% by the year 2000.

Kenya's extensive mangrove forests on the Indian Ocean coast are a major natural resource. They are spawning, nursery and feeding grounds for commercially important marine species like oysters, crabs, mullet and shrimp. They also serve as the habitat of particular species of birds. Furthermore, mangroves stabilize the shoreline with their extensive root system. Mangrove forests are harvested for poles used in Swahili-style housing and for boat-making. The forests are dense and contain as many as 20 different species of trees. Mangroves are halophytic plants which survive in their environment through a specially developed salt secretion process that makes them particularly vulnerable to coating by both inert and chemically active materials. Such materials carried to the river mouths are endangering the mangrove forests on the Kenyan coast.

Oil exploration in Kenya goes back to 1954. Nearly three-quarters of the country has been geologically mapped, with 30% being considered to have hydrocarbon properties, particularly in the eastern and coastal areas. Many wells have been drilled, several of them offshore, but no viable oil deposits have been discovered. Kenya imports an average of 3 million tons of crude oil from the Middle East annually and refines it at the East African Oil Refinery in Mombasa. This refinery serves the Ugandan market as well, and also exports some oil to the Far East. A pipeline links the refinery with Nairobi.

Kenya has considerable potential for hydroelectric power generation, although only a fraction of it is currently used. The Rift Valley has a large potential for geothermal electricity generation. At present, geothermal electricity is generated at Olkaria in the Rift Valley.

3.6 Djibouti

Djibouti occupies some 23,000 square kilometres of mostly arid land. Only about 6,000 hectares of land – in the north around Tadjoura, in the south around Ali Sabieh, and between Djibouti and Loyada – are considered arable land, even though precipitation does not exceed 200 mm annually, and less than 1,000 hectares are cultivated. Less than 10% of the territory is classified as rangeland, and there is hardly any woodland left. The residual forest in Day region north of Tadjura, believed to have covered 75 square kilometres two centuries ago, covers only a few square kilometres today. A few hundred hectares of irrigated land produce vegetables and fruit for the port town, using mostly water from wells. A rural development project launched by the government in 1981 to promote cultivation engaged some 1,000 people by the late 1980s.

Livestock raised by nomadic pastoralists is the country's main product, some of which is exported. A good portion of the livestock exported from Djibouti comes from Ethiopia and some from northern Somalia. Added to the domestic herds, this influx raises the stocking rates in Djibouti, which are believed to greatly exceed the carrying capacity of the land, with noticeable degradation as the result. Djibouti exports a number of agricultural products, including coffee, pulses, sugar and oil seeds, all of which are smuggled from neighbouring countries. Some fish is also exported.

Djibouti's economic reason for existence remains its port, which has long served Ethiopia via the 770 kilometre rail link to Addis Ababa. The port, which in recent years received an average of nearly 1,000 ships annually, represents the main source of state revenue and the main source of employment for the population, 60% of which reside in the town of Djibouti itself (Djibouti, 1990–91).

Water

Djibouti has no permanent bodies of fresh water in its territory. There are many seasonal streams (*wadis*) that flow intermittently towards the sea, but their potential has not been studied.

Minerals and energy

Djibouti lacks important mineral resources. There are salt deposits at Lake Asal, situated at 153 metres below sea level, believed to be the lowest point in the continent. There are limestone deposits that could support cement production, and perlyte, an insulation material used in construction. Djibouti is a region of active volcanoes and hot springs, and it is thought likely to have significant geothermal energy resources. The French located some geothermal sources in the 1960s and exploratory drilling is continuing, but exploitation has yet to begin. Djibouti's energy needs are met partly by fuelwood and charcoal and partly by imported fuels. Wood and charcoal are used in the rural areas and in Tadjura, where prices for these items are still modest. Kerosene is widely used in the towns, and petroleum is used to produce electricity.

4

Food Security

4.1 Overview

Food security means 'access by all people at all times to enough food for an active, healthy life' (World Bank, 1988, p. 1). Such access is not guaranteed to all people in the Horn. Indeed, food security is on a declining trend in all countries of the region. Evidence of the decline is an increase in national food deficits faster than the increase in the means to pay for food imports. The main food-insecure groups are resource-poor farmers, urban poor, poor pastoralists and refugees. More than a third of the region's population receive insufficient food to prevent stunting or to live a fully active working life. The major causes of food insecurity are poor overall economic performance, poor agricultural performance, wars, droughts and long-term environmental deterioration. These are the main findings of a study commissioned in 1989 by the Inter-Governmental Authority on Drought and Development (IGADD), which represents Ethiopia, Eritrea, Sudan, Somalia, Djibouti, Kenya and Uganda (see IGADD, 1990, p. i). This study put the proportion of 'food-insecure' people at that time in Somalia at 50%, Ethiopia 46%, Uganda 39%, Kenya 37%, Sudan 18% (1990, p. 14).

The problem of food security is a complex one. Famine has become its symbol in the Horn of Africa, but it is only one of its manifestations, albeit the most dramatic. Famine is an occasional and temporary visitation that results in high mortality rates due to starvation, whereas endemic under-nourishment – both in terms of inadequate calorie consumption and vitamin deficiency – is an enduring condition which results in high morbidity rates, stunting and shortening the lives of those who suffer from it. Focusing on famine tends to distort our perception of the problem, leaving the impression that its impact is indiscriminate and catholic. It is now understood, as Sen (1981) put it, that the characteristic feature of starvation is not that there is not enough food, but that some people do not have enough to eat. The list of causes of food insecurity listed in the IGADD report cited above omits the differentiating factor called 'entitlements' by Sen.

The issue of food security is confounded by the scarcity and unreliability of disaggregated statistical data. For example, figures of overall agricultural production levels reveal little about food availability and the ability to acquire it. In fact, in countries with large export cultivation sectors, rising production figures in the agricultural sector could mean reduced local food production, as more land and labour are devoted to export crops. The consequent loss of land and labour renders subsistence food production increasingly precarious and undermines food security in that sector. Figures cited below will illustrate this point.

The bulk of those who suffer food deprivation in the Horn are to be found in the rural sector, among subsistence cultivators and pastoralists whose production has been reduced because they no longer command adequate land and animal resources. Because they have been entirely deprived of these resources, refugees are another category susceptible to food deprivation; and here we should keep in mind that peasants and pastoralists comprise the bulk of the refugee population. In the main, the urban poor are refugees from rural poverty, and they lack the labour skills required to earn an adequate income. They accounted for an estimated 12% of the 'food-insecure' people in the Horn (IGADD, 1990). In both the rural and urban sectors, and in the refugee camps where they are the majority, the weaker members of the family – women and children – are in greater risk of food deprivation than men.

The direct cause of food insecurity is low and falling food output per capita, the result of stagnant or low agricultural production growth in the subsistence sector. Ethiopia is the worst case, where agricultural production increased by a mere 1% in the 1970s, and in the 1980s actually declined by an average 2.1% annually. The index of food production per head declined from 104.6 in 1960 (1979–81=100) to 86.4 in 1987, and the available food supply per head in grain equivalent fell from 174 kg in 1960 to 114 kg in 1987. (For details, see Pickett, 1991, pp. 188–189.) Daily average calorie intake in Ethiopia fell from 1,714 in 1969–71 to 1,476 in 1984–91 (FAO, 1991b). Sudan's rate of agricultural production growth was 2.9% annually during 1965–80, then fell to 2.7% during 1980–88. The average index of food production per capita fell to 89 in 1988 (1979–81=100) (World Bank, 1990, p. 184). The daily average calorie intake in Sudan fell from 2,213 in 1969–71 to 1,785 in 1991. Somalia's rate of agricultural growth during 1965–80 was an impressive 4.1% annually, but it fell to 2.7% in the 1980s. While per capita food production is said to have remained constant during the latter period (World Bank, 1990, p. 184), calorie intake in Somalia was reduced from 2,224 in 1970 to 1,336 in 1990. Kenya's rate of agricultural production growth during 1965–80 was even more impressive at 4.9% annually, but it fell to 3.3 in the 1980s. Per capita food production fell more than 10 points on the index scale of 100, and calorie intake went down from 2,249 in 1970 to 1,865 in 1991. Average annual growth rate in per capita food production for 1979–93 was reported for Ethiopia –1.2, Kenya –0.4, Somalia –6.0 and Sudan –2.2 (World Bank, 1994, 1995).

In countries where food production does not support large numbers of people, and where other sectors of the economy can provide the surplus needed to finance food imports, decline in food production need not result in food insecurity. The linkage between food production and food security is not close in such cases. However, in Africa generally, and the Horn particularly, food production is the main source of livelihood for the vast majority of the population, so even a small decline in production results in what Sen (1981, p. 48) calls 'a collapse in entitlements' – meaning loss of food as well as income for the mass of the population, for whom there is no way to make up the deficiency. Here the link between food production and food security is very close, because food production is in the hands of a mass of smallholders who are also self-providers. In Ethiopia, while agriculture contributes no more than 40% to GDP, it supports more than 80% of the population, and provides 90% by value of the exports. In Kenya, agriculture contributes less than 30% to GDP, but employs more than 80% of the labour force, and provides nearly all the exports. The corresponding figures for Somalia and Sudan are of the same order.

Inadequate food production in the Horn has resulted in rising food imports and an increasing dependence on external sources of food. Cereal grains and cereal-based products are the main food imports, led by wheat and rice. This is a sign of changing consumption patterns led by the urban sector. That trend leads away from locally produced grains, and has the effect of constraining market demand for locally produced food. Falling world grain prices in the 1970s and 1980s encouraged this trend. Locally produced coarse grains, such as sorghum in Sudan and white maize in Kenya, are occasionally exported, while refined grains are imported. By the end of the 1980s, all the countries of the Horn had food import dependency ratios, ranging from 6.2 for Kenya to 70 for Eritrea (Table 4.1).

Food import capacity depends largely on export revenues. None of the countries in the Horn realized increasing export revenues per capita in the 1980s, and the gap had to be filled with food aid. This was a period marked by drought and famine; in addition to imports, all countries received food aid from abroad. Indeed, the countries of the Horn were major recipients of world food aid throughout the 1980s. Estimates of the number of those in need of food aid in the late 1980s ranged up to one-third of the population of the region: more than 30 million people. In the early 1990s, Ethiopia, Eritrea, Somalia and Southern Sudan remained heavily dependent on food aid, while the situation in Kenya improved markedly. The enduring nature of the crisis led to the establishment of the United Nations Special Emergency Programme for the Horn of Africa (SEPHA). Food aid generally followed a rising trend even in non-drought periods, and had a similarly depressing effect on local production as food imports. Food aid was used to reduce the price of wheat flour and bread in the urban areas, while wheat and rice rapidly gained ground as inexpensive and convenient foods.

This chapter reviews each country's capacity to achieve food security, before examining the relationship between food security and regional trade. The only country not included in the review is Djibouti, which has no food production of its own and is entirely dependent on grain imports, mostly smuggled in from Somalia and Ethiopia. With three-quarters of its population living in towns, and a large European expatriate presence, Djibouti is heavily dependent also on food and drink imports from Europe.

Table 4.1   *Dependency on Foreign Sources for Staple Food as Percentage of Annual Food Production*

Country	Source	Dependency ratio	
		1981	1989
Djibouti	Food aid	31.6	20.5
	Import dependency	68.4	79.5
	Total dependency ratio	100.0	100.0
Eritrea	Food aid	–	35.0
	Import dependency	–	25.0
	Total dependency ratio	–	70.0
Ethiopia	Food aid	1.6	17.2
	Import dependency	1.8	8.6
	Total dependency ratio	3.4	25.8
Kenya	Food aid	6.8	3.0
	Import dependency	5.8	3.2
	Total dependency ratio	12.6	6.2
Somalia	Food aid	28.8	8.0
	Import dependency	37.6	20.3
	Total dependency ratio	66.4	28.3
Sudan	Food aid	4.2	7.3
	Import dependency	6.3	20.4
	Total dependency ratio	10.5	27.7

Sources: FAO Food Production Computer Print-Outs, June 1991; FAO Trade Yearbooks, Vols 37, 38, 40, 41; FAO, 1991. *Food Aid in Figures*, Vol. 8/1, Rome.

4.2 Ethiopia

The risk of food deprivation has been greatest in Ethiopia since the early 1970s, when the gap between food production increase and the rate of population growth began to widen alarmingly. This was hardly realized at the time, if only because the size of the population was grossly underestimated. Assuming an estimated rate of growth of 2.5%, the size of Ethiopia's population in 1970 was officially estimated at 24.3 million, whereas it was closer to 30 million (Pickett, 1991, p. 40). Nutrition surveys carried out in several regions before the 1972–74 famine showed very low levels of calorie intake, exacerbated by vitamin deficiency (Holt et al., 1975). Nor was this limited to the land-poor provinces of the north. A study of Sidamo Province in the south showed an average daily calorie consumption per adult male of 2,434, while the estimated daily requirement was around 3,500 calories (Sadler, 1976, p. 25). 'Deficit' areas are found on the eroded highlands of Tigray, Wolo, western Begemdir, northern Shoa, the eastern lowlands, eastern Hararghe, the Ogaden and the southern and western border regions. Many of these are pastoral areas.

Famine is no stranger to Ethiopia, especially to the northeast, where this one began (Pankhurst, 1985). According to one study (Mesfin Wolde Mariam, 1986, p. 147), in the two decades between 1958 and 1977, about one-fifth of Ethiopia's districts experienced famine conditions every year. The rainfall pattern became erratic as early as 1970, when the climatic trend that ravaged the Sahelian zone across the continent was felt in northeastern Ethiopia, mainly in the provinces of Wolo and Tigray, where yields were already half what they were elsewhere in Ethiopia. The peasants held on with the help of the annual small rains, but when these failed also in the spring of 1972, the people were doomed unless massive aid reached them in time. The failure of the rains and the policy of the imperial regime combined to produce a massive famine. No aid at all came until two years later, after famine had claimed the lives of an estimated 200,000 people.

The impact of drought was regional. According to a report by the Ethiopian Ministry of Agriculture (Ethiopia, 1973), only 21% of all districts had below normal production in 1972–73. Amartya Sen calculated the output for the entire country was only 7% below normal, 'hardly a devastating food availability decline' (Sen, 1981, p. 90). Taking Ethiopia as a whole, there need not have been famine in Wolo, Tigray and Hararghe, if surplus food had been diverted to the stricken region, and the people there had sufficient purchasing power. However, given the collapse of purchasing power in the northeast, market prices there did not rise enough to attract surplus food from other areas. The imperial regime not only did nothing itself to prevent the disaster, but succeeded for a time in concealing its existence, and rejected offers of assistance from abroad when it first became known. The claim 'drought is caused by nature, famine is caused by man' became the slogan of the revolution that ended the reign of the imperial regime in Ethiopia in 1974.

In the brave new world which dawned with the revolution, it was optimistically assumed such disasters would not occur again. Yet, far worse was to happen within a decade. The decline in food production per capita continued. The Ministry of Agriculture divided rural Ethiopia into high-potential surplus-producing zones and low-potential subsistence zones; the latter comprised two-thirds of all administrative districts in the country. Only two districts in the northeast (Wolo, Tigray) were in the first category. Low-potential zones had low priority in development assistance and agricultural inputs. Such support was concentrated in the surplus-producing areas of Gojjam, Shoa and Arsi. Estimates of per capita production of food under normal circumstances, produced by the Ethiopian Relief and Rehabilitation Commission, indicated that production in six out of fourteen provinces fell below the RRC's relief ration of 500 grams per person per day (Kumar, 1990, p. 178). In other words. many people in Ethiopia were starving in fair weather, even before drought unleashed another horrendous famine in 1983.

This time there was no attempt to conceal the famine, which again hit northern Ethiopia first, before spreading to many other parts of the country. Beginning in 1981, the RRC raised the alarm and sent out appeals for international aid. This had scant effect, and the military regime itself showed no concern, fully preoccupied as it was in waging civil war on several fronts. At the same time, the state-controlled media minimized the problem in order not to foul the atmosphere of the celebrations commemorating the regime's tenth anniversary, held in the autumn of 1984. By this time, the famine had affected eleven out of Ethiopia's fourteen provinces. It had become an international media event, and foreign aid was pouring in. Distribution encountered serious delays due to lack of transport, while the military regime refused to allow any of its own vehicles to be used for that purpose. It also refused to let aid cross the battlelines into areas held by opposition movements, where millions of people were trapped. People from Tigray, which was partly occupied by rebels, had to trek vast distances – up to 650 kilometres – to reach eastern Sudan and find relief. Ultimately, aid came too late for an estimated one million souls. It was only after the celebrations were safely over that the regime declared a national emergency.

It is fair to say that in Ethiopia 'famine is but an exaggeration of the normal circumstance in which domestic production and what can be had through external trade cannot provide all with enough to eat' (Pickett, 1991, p. 36). It is to be expected that production in some areas will fail at some time for some reason, be it drought, flood, locust, disease, armed conflict etc. Given the absence of reserves or other resources, famine can be averted only if food from surplus areas can be transferred to deficit areas in time. This is possible if there is some surplus in the country, as well as sufficient integration in the economy and the state apparatus to effect a transfer. Pickett points out that the size of the food surplus and the degree of integration are themselves dependent on the level of economic development: 'And the truth is that the Ethiopian economy has been neither sufficiently productive nor sufficiently well-integrated to deal effectively with serious failure of the rains' (Pickett, 1991, p. 36).

Pastoralists in areas affected by drought are the first to suffer the consequences and the last to receive aid. Since pastoralism is a resilient production system designed to cope with drought, it is a paradox that pastoralists should suffer more than others from it. However, as one study comments, 'while droughts are a normal occurrence in the dry regions of Africa, herders are confronted with processes of change that go well beyond the effects of climate' (Little, 1992, p. 1). He adds that while drought is a 'normal' state of affairs, the processes that have inhibited pastoralists' ability to cope with it are not. 'Coping strategies' for peasants and pastoralists alike in such times of stress are based mainly on movement. Peasants in Tigray, for instance, usually moved to less affected parts of their province where they have relatives, or sent some family members to work in the coffee-producing areas of southern Ethiopia and the mechanized farms in the Humera region of the west. The last two alternatives were denied them in the 1980s, due to the land reform which outlawed hired labour. As a final resort, they moved onto the main road artery that crosses their province, where they could at least be reached by relief operations. Before doing so, they had eaten the seed grain, sold their animals and agricultural implements, and even the wood of their huts to buy food.

Pastoralists have nothing to sell but their animals. As is well known, animal prices fall precipitously in such times, when supply greatly exceeds demand. When pastoralists sell animals to buy grain, the loss they incur due to the decline of exchange rates is catastrophic, as shown by Sen (1981) in his study of the Hararghe famine of the mid-1970s. To this must be added the loss of animals due to drought. Pastoralist coping strategies involve movement with their animals but, as described in Chapter 2, their movements have been greatly circumscribed in recent decades, and most of the riverine and other permanent water sanctuaries have been lost to commercial cultivation. The manifold conflict in Ethiopia added to the predicament of the pastoralists, who often found themselves trapped in war zones. Finally, the pastoralist habitat is the least accessible to relief operations, and starving herders have to trek the longest distances to find aid. Not surprisingly, one study (Maffi, 1975) shows the highest incidence of mortality during 1974–75 in Wolo was among the Afar pastoralists. Likewise, a study of Hararghe during 1974 (Rivers et al., 1976) shows the mortality rate for children under five years of age was two to three times higher for pastoralists in the Ogaden than for peasants on the eastern highlands.

Prospects for food security in Ethiopia are not bright. The estimated food deficit in the foreseeable future is in the range of 650,000 tons of grain per year, even if natural conditions remain favourable. The regime that came to power in 1991 has put its hopes on expanding the area under rainfed cultivation and the promotion of small-scale irrigation schemes based on water harvesting and conservation. It is also promoting the use of inputs such as fertilizer. It would require an annual food production increase of 5% for the country to reach self-sufficiency in the near future. Clearly, this is a target not likely to be reached, and Ethiopia is unlikely to attain food security in the foreseeable future.

4.3 Eritrea

When it was a province of Ethiopia (1952–91), Eritrea was among the most 'food-insecure' regions in that country. It did not have a single high-potential surplus-producing district. Per capita food production in a normal year during the early 1980s was 247 grams per day, the second lowest in Ethiopia, and far below the absolute minimum figure of 400 grams. Eritrea was hard hit by drought in the early 1980s. A series of bad harvests resulted in severe food shortages that culminated in massive famine in 1983. Food output that year was estimated at 45% below normal, and the situation grew worse when the 1984 harvest was also lost due to the poor rains. Eritrea had been in the grip of war since the early 1960s, and its productive capacity was seriously undermined by the conflict. The war caused massive and prolonged destruction of plant and animal life systems, and accounts for the extreme dependency into which Eritrea has fallen, as well as for the lack of reliable statistics. The war also prevented food aid from reaching the population who lived in areas controlled by the nationalist liberation movement. Like the people of Tigray, Eritreans had to trek vast distances and cross the Sudanese border in search of relief.

No reliable data exist on agricultural production in Eritrea (Wolde Amlak Araia et al., 1994). The Ethiopian Statistical Authority stopped collecting production data in Eritrea long ago due to the war in that area. Yields are known to be the lowest in the highlands. According to a 1949 FAO report, Eritrea's yield in cereals and pulses per hectare was only 0.5 metric tons, compared to 0.8–1 in the Middle East, and 2.3 metric tons in Egypt at the time. The data cited below come from a 1987 FAO Crop Assessment Mission (FAO, 1987), and a study done also in 1987 for the Eritrean Relief Association, an agency of the Eritrean Peoples Liberation Front (EPLF/ERA, 1987). Eritrea's annual food requirements are estimated at 400,000 tons; under good rainfall conditions, Eritrea can produce between 225 and 250 thousand tons, or 55–65% of its needs. Production during the war years varied greatly, ranging from 40,000 tons in 1987 to 170,000 tons annually during 1980–83.

4.4 Sudan

The national rainfall series for the countries of the Sahel (IUCN, 1989, p. 8) show that Sudan had a persistent precipitation deficit in the three decades from 1960 to 1990. The central riverine region is least affected by national rainfall patterns because cultivation there depends on irrigation, not precipitation. It is the regions to the west and east of the Nile in Northern Sudan, where subsistence cultivation and pastoralism are the main modes of production, that are most vulnerable to climatic vagaries.

Drought is no stranger to these regions, and the people evolved various ways of dealing with its visitations. To cope with different rain and temperature conditions, they planted a range of crops to ensure that some would survive. While this pattern of risk aversion reduced yields in good rainfall years, it also provided food protection in dry years. The trend towards monocropping of millet in recent years eliminated this safeguard. In the past, when almost all production had been for subsistence, cultivators in the west would store the occasional surplus as protection for short-term food shortages. This practice was abandoned when cash crops were introduced with the encouragement of the state, and people reduced their production of staple food grains in order to produce sesame, groundnuts etc. The cash they earned was used to buy clothes, tea, sugar, as well as food grains like sorghum (Mustafa, 1991). To deal with famine conditions when they occurred, people relied on a variety of 'coping strategies'. These included reliance on the earnings of migrant labour, sales of assets, consumption of wild foodstuffs and, ultimately, migrating temporarily to other areas (de Waal, 1990).

Drought hit the western Sudan in the early 1970s and caused severe damage. Known by the name *Ifza Una* ('save us'), it killed a large number of acacia trees and disrupted the production of gum arabic, reducing incomes as well as food availability. Nevertheless, it did not unleash a famine because traditional 'coping strategies' were still effective, one of them being to move southwards to areas less affected by drought. This was a time of peace, and such areas were not only accessible but also produced a surplus of food that was marketed in food-deficit areas. Drought hit the eastern region of Northern Sudan in the second half of the 1970s. It caused heavy losses of livestock, and many impoverished Beja pastoralists were forced to abandon their traditional way of life (Dahl, 1991). Food shortages and starvation among the pastoralists began as early as 1975. In similar circumstances in the past, the Beja had found refuge with their animals in the Gash and Tokar inland delta areas, and even moved onto the foothills of the Eritrean plateau. Now the river deltas were taken up with cotton and castor-bean cultivation, while war in Eritrea closed that avenue of escape.

Famine on a massive scale was unleashed in the east and west when, after several years of inadequate rainfall, the rains failed completely in 1984. Livestock was wiped out in both regions. Because of the resumption of the civil war, pastoralists from Kordofan and Darfur were prevented from moving southwards, as they used to do in the past. For the Beja, mobility had become even more constricted by the presence of large numbers of Beni Amer pastoralists who had moved into eastern Sudan to escape the war in Eritrea.

When people began to starve in Darfur, their only option was to move to the central riverine region, an immense distance. In the east, many Beja starved in their isolated hamlets before the survivors sought refuge in the camps already set up to provide relief for the victims of famine in Tigray and Eritrea. Huge numbers of people, particularly from the far west, were on the move, trying to escape the famine in Sudan in 1984. Their plight went unnoticed, because the regime in Khartoum ignored reports and pleas from the famine-stricken areas, and the world's attention was focused on the tragedy in Ethiopia. It was not until the end of 1984, when starving westerners crowded into the outskirts of Omdurman and the Beja survivors appeared in the refugee camps, that wider awareness of the Sudanese tragedy dawned.

The regime's reaction was one of embarrassment and anxiety to minimize the political damage at home and abroad. It ordered the refugees from the west to return to their homeland where, given the lack of transport, relief was unlikely to reach them. When they refused, soldiers forcibly loaded them on lorries for transportation to Darfur. Famine relief was left to foreign aid organizations, whose resources were badly strained by the famine in Ethiopia.

Drought persisted through 1985. An unknown number of people died of starvation during that year: not because food was not available in the country, but because it could not reach them in time. The main obstacle was the sorry state of the railway and roads leading west.

The eastern and western regions of Northern Sudan remain areas of chronic food insecurity, and famine is a constant threat there. Drought has visited Northern Sudan intermittently, and the conditions that turn drought into famine have not improved; indeed, they have worsened. Despite the low population density in the west, land shortages have affected the subsistence sector there. It is estimated that a family of six requires 15 hectares of land to raise enough grain for its own consumption, but few families have enough land to devote 15 hectares solely for raising grain. Productivity in the rural sector has continued to decline. According to an official study published in 1977, sesame yields in Kordofan were 14 times higher in 1961 than in 1973. For *dura* and *dukhn*, yields were four and three times higher respectively in 1961. According to another study, sorghum yields in the peasant sector fell as much as 43% between 1960 and 1985 (Mohamed Salih, 1987). There are wide fluctuations in production from year to year, particularly in the subsistence sector, due to the erratic pattern of precipitation. For example, the production estimate in this sector for 1993–94 was 48% lower than the previous year due to low rainfall, late planting, and the presence of pests (FAO, 1993b).

The Sudanese economy in general has experienced a steady decline for nearly two decades, and its ability to finance imports has diminished accordingly. In order to gain foreign exchange, Sudan abandoned an earlier policy of maintaining 'strategic sorghum reserves' at the national and provincial levels. Sudan exported sorghum to Saudi Arabia, enticed by a generous subsidy offered by the importing country, which uses sorghum as stockfeed to build up its own animal production. Sudan's infrastructure, particularly the transport system, is in a state of advanced decay. Moreover, the civil war, which is largely responsible for the economic plight of the country, intensified in the early 1990s and spread to more areas, like the west of Darfur and the Nuba Mountains. Famine conditions were reported in the west during the late 1980s and again in the early 1990s.

Among those at risk of food deprivation were the displaced people who congregated in the region of the Three Towns (Khartoum, Khartoum North, Omdurman) since the famine of the mid-1980s. Many of those who made the trek from the far west to escape starvation never returned to their homeland, finding refuge in the sprawling slums that mushroomed around Sudan's principal towns. Southerners fleeing the horrors of renewed warfare in their region began to arrive at about the same time, and their numbers were soon to reach the hundreds of thousands. Many of these internally displaced people depended for their survival on food aid provided by foreign organizations, and their living conditions were atrocious. Furthermore, the Sudanese authorities adopted a policy of forcefully removing displaced people from the region of the capital, and transporting them to distant locations in the interior of the country, where they were more or less left to fend for themselves.

The civil war is directly responsible for turning Southern Sudan into a chronically food-insecure area, where famine has taken a steady toll of human lives for more than a decade. With land, water, livestock and fish in abundance, and its traditional food production systems undisturbed by commercialization, this region is self-sufficient in food under normal conditions. Unfortunately, as described in Chapter 5, conditions have not been normal in Southern Sudan for a very long time, and the region has known precious few years of peace during the second half of the 20th century. Southern Sudan has been the battleground of a many-sided conflict, involving not only the main struggle between the North and South over state power, but also a struggle among several Southern factions for control of the Southern resistance movement, as well as numerous local conflicts over resources fought among ethnic groups in the South.

The result was a proliferation of armed groups which, to varying degrees, depended for their food upon the local population. This they got mostly by looting, or, as in the case of the Sudan Peoples Liberation Army (SPLA), by imposing arbitrary taxation. Since the production system in the South was designed for subsistence and not for surplus, the expropriated food represented a net deduction from local consumption. Even worse was the policy of deliberate dislocation and destruction of production, intended to deny food to the enemy. This was widely applied by the Sudanese government forces against the SPLA, by the SPLA against rival Southern factions, and by the various factions of the SPLA against each other. People were removed from their villages and herded into 'strategic hamlets' where congestion soon decimated the herds. Their homes and crops in the fields were burnt, their water sources poisoned, and their animals killed (Human Rights Watch/Africa, 1994).

A massive dislocation of the Southern economy, an equally massive displacement of its population, and a universal state of extreme destitution have been the results in Southern Sudan. Violence and famine have been stalking the region for years now, forcing many people to seek refuge in Northern Sudan and in neighbouring countries. In the late 1980s, famine began to take an increasing toll in the South, and world attention was drawn to yet another tragedy in the Horn of Africa; the worst year was 1991. Efforts by numerous international aid organizations to reach the hapless victims of war in Southern Sudan were hampered by the isolation of that region, the nearly total lack of transport and communication links, and the chaotic conditions prevailing there because of the manifold conflict. However, the most serious obstacle proved to be the attitude of the warring parties, which sought to prevent food aid from reaching areas controlled by the enemy. For example, the SPLA obstructed efforts to bring food to Juba, the main town of the region, where large numbers of people had congregated and were slowly starving – because it was under government control. The Sudanese regime, on the other hand, obstructed efforts to bring food aid to areas controlled by the SPLA (de Waal, 1993).

4.5 Somalia

Conflict is primarily responsible also for making Somalia a ward of international charity in the 1990s. The country's chief economic activity is nomadic
pastoralism, practised by more than half the population. It has provided
employment opportunities for many more, contributed more than one-third
of the GDP, and accounted for up to 80% of the value of exports. Nomadic
pastoralism is a resilient production system designed to withstand adversity
of all kinds, including drought. However, the drought that lasted from early
1973 to mid-1975 in Somalia, where it is recalled as *Dabadheer* ('long-tailed
drought'), had a devastating impact on Somali herders, especially in the
northern region. The drought claimed nearly half the country's flock animals
and a third of its cattle. Hundreds of thousands of people streamed into the
towns in search of water and food, and close to 300,000 ended up in some 20
refugee camps. By 1975, the famine had killed 20,000 people. Drought
recurred in 1979–80, at a time when three-quarters of a million war refugees
entered Somalia from Ethiopia.

The pastoralist sector recovered quickly. Exports of flock animals and cattle
actually increased and exceeded pre-drought levels, until a ban imposed by
Saudi Arabia in 1983 on livestock imports from Somalia for health reasons
brought an end to the boom in this trade. However, it had been noticed before
1983 that, while the number of animals marketed increased, the average weight
per livestock began to decline in the late 1970s (ILO, 1989, p. 38). This was a sign
of the distorting influence of the market on a mode of production not designed
for trade. The technology and organization of livestock production in Somalia
had not changed essentially from the days when it served subsistence needs.
Reflecting recently acquired cash needs of the nomads, the lure of the market
was changing their economic behaviour, not unlike the case of subsistence peasants who were persuaded to reduce the area planted for food in favour of cash
crops. The state encouraged this trend by improving water supplies, stock routes,
veterinary services and marketing facilities. In responding to market demand by
over-stocking, the pastoralists were in fact exhausting the land, the major asset
of their production system, and degrading its product, the livestock.

Agriculture has traditionally occupied some 20% of the population in
Somalia, and accounted for about 7% of GDP. Grain production was heavily
concentrated in the southern part of the country, especially the Juba–Shebeli
region, which also produced all the rice. The central and northern regions
were grain-deficit areas, except for the Gebile district in the extreme northwest. During the 1970s, grain production was stagnant, while banana and
sugar production declined. Liberalization of state policy during the 1980s
resulted in a steady increase of the area under cultivation as well as output.
There were impressive gains in maize and sorghum production, so that by
1985, according to official calculations, Somalia was approaching self-sufficiency in staple foodstuff (Somalia, 1987, p. 152). Food-grain imports were
reduced from 73% of total grain in 1982 to 41% in 1985. Prospects for
achieving food security seemed good.

These prospects were destroyed by the conflict that engulfed Somalia in second half of the 1980s (Mubarak, 1996). Initially the conflict concentrated in the north, where it devastated the regional economy. Towns were reduced to rubble, the infrastructure demolished, thousands of people perished, and the survivors with the remaining livestock fled across the Ethiopian border into the Ogaden, where many spent years in wretched refugee camps. At the turn of the decade, the conflict spread to southern Somalia. There it attained epidemic proportions, until the state collapsed completely and the country was divided into clan fiefdoms. With the cessation of production and the breakdown of the market, famine ensued in many parts of Somalia. Food became a weapon in the civil war, and many lives were lost, not because food was not available, but because it could not be brought to those who needed it. It required an armed intervention led by the United States and under the aegis of the United Nations in 1992 to bring food to the starving people of Somalia. The intervention exacerbated political strife, forcing the withdrawal of the United States troops and the suspension of most United Nations' operations in Somalia. A number of nongovernmental organizations remained in the country and continued to channel aid to its people.

4.6 Kenya

Although Kenya was spared the extreme experience of the other countries in the Horn, large sections of its population lack food security, and the outlook for the country's future in this respect is not bright. The last year when domestic food production satisfied national requirements was 1978, and the country has since become increasingly dependent on food imports. Serious food shortages appeared for the first time in 1980–81, when emergency food imports were required to avert famine. These shortages were seen as the result of farmers' negative reactions to the short-sighted policies of the National Cereals and Produce Board. By the time these policies were revised, severe drought hit Kenya in 1984 and famine threatened many regions of the country. Maize production fell by 35% and wheat by 43%, and livestock suffered considerable loss. Disaster on the scale experienced at that same time elsewhere in the Horn was averted by massive food aid from abroad as well as purchases of food, and an efficient distribution system that reached the needy areas in time.

Coffee and tea, Kenya's main export crops, were less affected by drought. They continued to bring in foreign exchange, some of which was spent on food imports. A 'food-crop versus cash-crop' debate ensued. One side decried the preference given to cash crops in view of increasing deficiency in food production, and the other maintained export crops could provide the money required for food imports while, at the same time, being more labour-intensive, the cash-crop sector could provide employment for the country's burgeoning population. The government itself believed there was a good balance between food and cash crops. Maize production already occupied nearly half the land under cultivation and could not expand farther. Production of higher value crops could not be reduced without a sharp fall in the per capita value of agricultural production, and loss of badly needed foreign exchange. Increased food production had to be achieved through higher productivity, with the assistance of more and better inputs. *The Development Plan, 1989–1993* (Kenya, 1989) envisaged a variety of strategies intended to raise production of both food and cash crops.

In the 1960s and 1970s, agricultural production in Kenya was boosted by expansion as well as intensification. The spatial limits of expansion were reached in the 1980s, with only semi-arid and arid regions remaining, where cultivation without irrigation is a hazardous enterprise at best. The cost of irrigation – something like USD 7,000 per hectare – is prohibitive, and that option has been abandoned. In order to attain self-sufficiency in food production, Kenya's agricultural output will have to expand at a rate that few developing countries have reached on a sustained basis. Raising yields by using more and better inputs, the provision of services, and incentives such as higher farmgate prices – the option chosen by the government – confronts severe constraints related to the high rate of population growth.

Despite the degree of commercialization already attained, agricultural production in Kenya is still dominated by smallholders. They produce 70% of the maize, 65% of the coffee, 50% of the tea, and nearly all the rice, pulses, cotton and pyrethrum. Horticulture is the only sector dominated by large farms. Of the 2.7 million smallholders in Kenya – of whom more than one-third are women – 80% have less than 2 hectares of land. Rapid population growth is forcing the pace of landholding fragmentation, a process which is bound at some point to obstruct the type of intensification envisaged by the government.

Although high for the region, the rate of production growth throughout the 1980s and early 1990s remained well below the rate of population growth, confronting Kenya with the same dilemma facing all countries in the Horn (Heyer, 1990). Agricultural production declined for four years (1990–93) because of drought, and food imports accounted for 10% of the total import expenditure in 1994 (Kenya, 1995, p. 99). Kenya has a high incidence of malnutrition, especially among children, as shown in the Household Food Security and Nutrition Surveys carried out since 1977 (see Kenya, 1990). Up to 35% of the children measured in 1994 were moderately or severely stunted. Among the vulnerable groups are the urban poor. While Kenya's record of industrial development is outstanding by regional and African standards, it has fallen far behind the country's job-creation requirements. There were fewer than 300,000 people employed in industry and trade by the beginning of the 1990s, and there was massive unemployment among the young urban population.

Communities in the semi-arid and arid regions are most at risk of food deprivation. They include farmers on marginal lands who practise high-risk cultivation, and traditional pastoralists who have become marginalized in

Kenya's modernizing economy. Pastoralists in the northern lowland region have long been afflicted by war, environmental decline and frequent drought, and many communities have become quite dependent on foreign assistance for their survival. For example, the Turkana have been supported by a food-for-work programme ever since the beginning of the 1980s. Food aid has become an essential component of an impoverished pastoralist society, which exists with minimal livestock holdings that could not sustain a viable pastoralist economy.

4.7 Constraints on Regional Food Trade*

There are several problems with food production and trade in the Horn. The first is the fragile production base. The principal crops produced in the Horn are maize, sorghum, millet, wheat, *teff* and barley. All the countries, except Djibouti, produce these crops – apart from *teff* which is grown only in Ethiopia (Table 4.2). The most important crops produced in these countries have no active demand in world trade, except for coffee, tea, livestock and some lowland crops like sesame and oil crops. Thus many principal crops are not potential vehicles of development for these countries. There is little regional trade because the major deficit areas are too far from potential markets in neighbouring countries, although there are exceptions to this rule, as in the case of surplus sorghum-producing areas in eastern Sudan adjacent to deficit areas in northwestern Ethiopia and western Eritrea. From a global as well as regional point of view, the crop mix makes for a narrow-based agro-economy that requires adjustment and diversification in the long run. Moreover, aside from pulses and sorghum, most crops produced in the Horn have limited nutritional value.

Nowhere in the world is agriculture as volatile as in the Horn. Although it is difficult to calculate variability indices, some general conclusions can be drawn from Table 4.2 and studies done for IGADD. Under normal conditions of production, the terms 'surplus' and 'deficit' tend to lose their sharp demarcation. In 1986–87, Tigray in northern Ethiopia produced a 'surplus' of cereals and was a major supplier for Eritrea. This was mainly because the province had good rainfall that year, in contrast to what had happened two years earlier when the rains failed and hundreds of thousands died of famine. In the context of the market, these terms are also untenable. 'Terminal markets' like Addis Ababa, Khartoum, Mogadisho and Nairobi, with no production of their own, can acquire a surplus, while areas where this surplus comes from can revert to a position of deficit in a bad season, and will then require imports from the capitals. In this sense, 'surplus' and 'deficit' are transitory terms, related to different sets of costs and prices. In nutritional terms, surplus areas may have serious malnutrition problems because of poor

*Sections 4.7 and 4.8 are abridged versions of Mulugeta Bezabeh's contribution on 'A Free Flow of Food Across Borders in the Horn of Africa', presented at the PRIO/UNEP seminar in Sigtuna, Sweden, September 1991.

Table 4.2 *Food Production in the Horn of Africa by Important Categories (1000 metric tons)*

Country	Commodity	1980	1985	1990
Ethiopia	Cereals	5,612	4,820	6,297
	Roots & tubers	1,632	1,651	1,718
	Pulses	919	558	38
	Sesame seed	37	37	599
	Meat	528	573	–
	Sugar cane	1,404	1,600	1,720
	Coffee	187	170	195
	Vegetables	498	550	595
	Fruits	203	215	228
Kenya	Cereals	2,233	2,901	2,729
	Roots & tubers	1,191	1,615	1,512
	Pulses	121	177	211
	Sesame seed	8	7	8
	Meat	286	314	389
	Sugar cane	4,532	4,023	4,750
	Coffee	91	94	104
	Vegetables	483	502	631
	Fruits	650	724	879
Somalia	Cereals	267	514	581
	Roots & tubers	39	44	50
	Pulses	9	24	14
	Sesame seed	38	57	50
	Meat	134	146	169
	Sugar cane	20	444	240
	Coffee	–	–	–
	Vegetables	27	42	57
	Fruits	18	19	45
Sudan	Cereals	2,843	4,069	2,059
	Roots & tubers	293	243	152
	Pulses	55	60	58
	Sesame seed	221	131	66
	Meat	477	441	395
	Sugar cane	1,283	4,800	4,300
	Coffee	–	–	–
	Vegetables	772	890	930
	Fruits	94	125	127

Sources: FAO Food Production Computer Print-Outs, 1991.

purchasing power among a segment of the population. By contrast, a deficit area may have a higher average per capita calorie intake than a surplus area, due to resource use, consumption habits, cultural practices in food preparation, etc. For these reasons, and also because production statistics of the countries under review are not disaggregated into surplus and deficit, these terms should be used with caution.

Economic analysts have advanced many reasons to explain the stagnation of production in this region. Wars and civil unrest are often cited as major reasons. In fact, peasants, given good rains, have always managed to plant on time, and harvest, store and market their produce, even in time of war. Other analysts cite environmental degradation and unfavourable climatic trends: yet environmental degradation is an age-old process, not a sudden change, and unpredictable weather is not peculiar to the Horn. Still others cite the inefficiency of the marketing system as a constraint on production: yet the Horn has been an active trading region throughout its history.

The root cause of stagnation has been the policy followed by governments in the region. As an illustration we may take a specific incident when policy-makers intentionally killed a mushrooming production system in a lowland area near Ethiopia's northwestern borders with Sudan and Eritrea. Though it has little rainfall, this area contains close to one million hectares of fertile topsoil. In the mid-1960s, a handful of farmers from Eritrea started mechanized farming, producing sesame and sorghum, and Ethiopia was soon to become an exporter of sesame. In less than five years, production amounted to 60 thousand tons, most of which was exported.

Interest in the area grew immensely, and external funding for infrastructural activities became readily available. In 1971–72, a massive capacity-building exercise with inter-disciplinary technical studies touching all areas of development – soils, water, transport, agronomic trials, storage, marketing, etc. – was undertaken with World Bank financing. A great deal was learned from the ingenious manner in which mechanized farming operated in this fragile environment of low rainfall, and particularly how farmers prepared themselves for the crucial period of torrential rains, during which they harrowed (not ploughed) the sandy soils with giant harrows, followed immediately with seeding. A remarkable capitalist enterprise in this formerly deserted area provided seasonal employment for more than 100,000 people from nearby Tigray and Gondar.

The Ethiopian revolution of 1974 dealt a mortal blow to this enterprise. In March 1975, the military regime that overthrew Emperor Haile Selassie decided to nationalize commercial farming. Commercial agriculture had made impressive strides in the Awash Valley and in the northwest, and was the only sector in agriculture that produced appreciable surplus. The regime intended to settle impoverished peasants from the north in the northwestern sesame-producing area, while the cotton-producing lands on the Awash were to be turned over to a state corporation. There were technical arguments against this policy, but the regime had an ideological commitment against emerging capitalism. Four months later, the Ethiopian farmers who, with their 750 tractors, had been producing sesame in the northwest, moved across the border into Sudan and established new farms there. The military regime tried to set up state farms in the northwestern lowlands, only to withdraw after two harvests with a net loss of USD 20 million. Afterwards, the area became a wasteland.

4.8 The Economics of Regional Food Trade

Three fundamental and closely interrelated economic issues determine the flow of food from one area/country to another: pricing and exchange rates, intra-regional trade behaviour, and informal sector trade. A review of these issues is important in order to identify the potential for food trade across borders, the constraints in achieving the potential of such trade, and the areas of cooperation required to achieve a sustainable food flow. Before embarking on an analysis of these issues, it is essential to take an overall look at the marketing structure of the countries in question.

Throughout the Horn of Africa, the effectiveness of the marketing system has been compromised by various factors – administrative and regulatory intervention, poor market structures, high cost of internal marketing etc. – rendering the market unable to arbitrate prices between locations over time. Even though administrative restrictions on internal marketing activity are on the way out, the other factors indicated above, including financial and managerial constraints, severely limit the effectiveness of parastatal marketing agencies in the supply or price control role. Public and/or private sector marketing cannot yet effectively fulfil distributive or cumulative functions.

As to pricing and exchange rates, several studies indicate a wide variety in pricing behaviour. This supports the conclusion that there is significant unexploited trade potential within and between the countries of the Horn. There is considerable variation in market prices between market/production locations within each country, even though all countries have pan-territorial official pricing policies. Price variation has been particularly marked in Ethiopia, an indication of how ineffectively marketing policies have permeated the economy. A perennial 'deficit' production condition in the north has accentuated this situation. The same applies to Eritrea, with high prices at the centre and relatively lower prices in the western lowlands. Price variation within Eritrea, however, has been less significant than variation between Addis Ababa and Asmara. In Somalia, price variations have generally been less marked, with greater variation in the northerly areas. In Sudan there has been a marked price gradient reflected in lower prices for sorghum in the eastern region around Gedaref, the major sorghum producing area. Primarily because of high internal transport costs, prices have tended to rise with movement towards the deficit western region of Sudan.

Seasonal market price variation has been high in all countries of the Horn, ranging from as much as 150% in Eritrea, 53% in Ethiopia, 60% in Somalia, to 25% in Sudan (IGADD, 1990). Such seasonal variation indicates ineffective market structures. Although all countries used to have official purchase prices supposed to be pan-seasonal, they were either too limited or too low to have a stabilizing impact. Seasonal price variations between countries in the Horn show a significant difference between seasonal price peaks and low points, especially noticeable in border production zones. A good example is the low price (across seasons) for sorghum in Gedaref, as opposed to higher prices for the same product on the Ethiopia–Eritrea border. A similar situation existed for sorghum prices in lower Shebeli inside Somalia, vis-a-vis relatively higher prices across the border in the Ogaden, inside Ethiopia.

Before drawing conclusions regarding policy implications, we still need to examine the nature of producer and market prices, and how these prices are positively or negatively affected by the official or parallel market exchange rates. We shall then compare the relative price ratios and the absolute level of prices with prices of comparable food commodities on the world market. Only by relating regional price ratios and levels to world levels can we determine the sustainability of intra-regional food trade.

In the 1980s, official purchase prices – referring to parastatal prices – were lower than market producer prices in Ethiopia and Somalia for all crops. Only in Somalia were parastatal purchases for maize and sorghum well above market prices, which reflects another weakness of highly subsidized purchases. Obviously, the parastatals do not play a balanced role as buyers of last resort. Second, official price ratios for Ethiopia and Sudan, using official exchange rates, were far higher than world price ratios. This is an indication of several possible factors, including higher production costs and over-valued currencies, the latter being the more plausible explanation. For Somalia, the opposite was the case. Measured at official exchange rates, official parastatal maize and sorghum purchase prices were well in excess of world prices. However, using parallel exchange rates, we find parastatal prices lowest in Ethiopia. Market price ratios appear to approximate world price ratios in Somalia because that country had a liberal food import policy; also due to its geographical position, it was more integrated into the world market.

The conclusion to be drawn for all countries of the Horn on prices in relation to exchange rates is that, at official exchange rates, market prices of basic food commodities were high compared to world prices. However, measured at the parallel market exchange rate, the absolute level of prices in the Horn in general appears to be near world prices, FOB East Africa (see IGADD, 1989).

The second key issue is intra-regional trade. This is insignificant in the Horn, although livestock trade, especially the unofficial one, is believed to be considerable. The average annual cereal trade in the 1980s was estimated at 70,000 metric tons, half of which comprised re-exports of rice from Djibouti. IGADD's *Study of the Potential for Intra-Regional Trade in Cereals in the IGADD Region* (1989) came to the conclusion that there is scope for intra-regional trade, although serious constraints will have to be overcome. Among these are the orientation of states towards self-sufficiency in staple food, over-valued official exchange rates which place commodities above world prices, lack of effective demand, stagnation in production, unrestricted food aid, non-convertibility of local currencies, difficult payment procedures, lack of positive policy concerning unofficial cross-border food trade, lack of market information and inadequate infrastructure.

The third key issue is unofficial trade. By conservative estimates, unofficial food trade is three to four times larger than official trade. Its very nature makes the extent of such trade difficult to measure. Various factors determine the existence and scope of unofficial markets. Among these are differences in comparative advantage between adjacent areas, seasonal price variations across borders, shortage of consumer goods on one side of the border, differences in economic support for commodities, absence of processing facilities on one side, etc. While governments recognize that unofficial trade has taken over the economy, they have been unable to devise sound economic policies and institutions to integrate the unofficial and official sectors of the market. Instead, they rely on administrative measures, such as border control, that have proved not only ineffective but indeed counter-productive.

PART TWO
CONFLICT

Throughout the second half of the 20th century, the Horn of Africa has been the scene of violent conflict occurring in all the countries of the region and involving most groups living in it. The conflict has undermined the national economies, bankrupted the states, made a mockery of development plans, and reduced the mass of the people to extreme poverty. The phenomenon is usually dubbed 'ethnic conflict', a label that carries its own facile explanation and discourages analysis of objective causes. There is no doubt that ethnicity – political mobilization on the basis of a collective identity based on cultural affinity – is frequently, though not always, one variable involved in the chemistry of the conflict. However, ethnicity is often perceived as the independent variable and determining factor in the generation of conflict, and that is a misperception of reality. Since the assertion of ethnic identity and aspirations does not always attain political expression, we need to inquire into the circumstances that encourage the politicization of ethnicity and lead to 'ethnic' conflict.

Democracy has not flourished in the post-colonial state in Africa. In fact, the monopolization of state power on the part of some groups is the rule rather than the exception. In the Horn, this monopolization took what Mazrui (1975) called an 'ethnocratic' form: that is, the monopolization of power by one or more ethnic groups and the consequent exclusion of others. The Ethiopian Empire, as it was called until 1974, was the creation of the Amhara, the major branch of the Abyssinian family, who retained exclusive control of the multi-ethnic state until very recently. Likewise, Sudan, with its melange of ethnic groups, has been controlled since its independence by the Arabs of the North. Neither the Amhara nor the Arab Sudanese are a majority of the population in their respective countries.

In both these states, the ruling ethnic groups sought to impose their own culture on the rest, in the guise of 'national integration'. Thus, Abyssinian Christianity and the Amharic language became the twin hallmarks of Ethiopian state nationalism, while Islam and Arabic became the symbols of official Sudanese nationalism. The cultural heritage of other groups was ignored. In Ethiopia, no language other than Amharic (and English) was allowed to be taught, printed, broadcast or spoken at public functions. Islam, the religion of a great many Ethiopians, was ignored, and the teaching of Arabic, essential to Muslim religious practice, was severely restricted. In Sudan, it was Christianity that faced serious handicaps, and it was the Southern Sudanese converts to this faith who suffered the consequences. Thus, it was the state in the first instance that politicized ethnicity through a policy of cultural oppression, thereby encouraging the assertion of cultural identities on the part of oppressed groups in their struggle for political emancipation. Inevitably, ethnicity emerged as the ideology of political conflict in many instances. Like all ideologies, however, it is not the cause, but the symptom of social disorder.

The conflict is political because the bone of contention is state power. Monopolization meant that members of excluded ethnic groups lacked access to state power. In Africa, where the state controls both the production and distribution of material and social resources, exclusion from state power is tantamount to material and social deprivation. In the Horn, this reached exaggerated proportions under the self-styled Marxist military regimes that ruled Sudan, Somalia and Ethiopia during the 1970s and 1980s, when the state acquired total control of the economy and the national assets. In those circumstances, it became imperative to struggle for state power in order to acquire a share of resources, especially for regions and ethnic groups already deprived due to the prevailing uneven pattern of development.

Regional disparity in material and social resources in the Horn is the result of uneven natural endowment and historical processes, both described in the preceding chapters. Where nature has been meanest, as in the case of the arid lowlands, or renders a region inaccessible, as in the Southern Sudan, or has depleted its original endowment, as in the northern plateau, there was little opportunity for development during the colonial era; except, of course, where irrigation made it possible. The pattern of economic development did not change essentially in the post-colonial period. As a result, many regions saw little sign of development before and after independence, and their inhabitants were bereft of the economic and social resources that became available in other, more fortunate regions. The starkest contrasts in this respect are (a) between the arid regions inhabited by pastoralists throughout the Horn and areas where commercial cultivation brought a measure of prosperity; (b) between the riverine region of Northern Sudan and the eastern, western and southern regions of that country; (c) between northern and southern Somalia; (d) between the central and southern parts of the Ethiopian plateau and the heavily eroded northern region.

In the Horn, the regions where conflict has flourished are also those most deprived by nature and the historical processes mentioned above: northern Kenya, Southern Sudan, Darfur, northern Somalia, Djibouti, Danakil, Ogaden, Bale, southern Sidamo, Tigray and Eritrea. Eritrea is often regarded as an exception, because in the early postwar years its economy was more developed than that of Ethiopia. This ignores the fact that development in Eritrea was limited to the central highlands where Christian cultivators predominate, while the conflict commenced in the western lowlands where Muslim pastoralists live. Moreover, by the time Christians began to join the nationalist movement, the Eritrean economy had faltered. These regions are inhabited by ethnic groups that were not part of the ruling strata in their countries; consequently, they lacked access to state power. Political impotence and impoverishment are closely linked and mutually reinforcing handicaps. In order to redress material and social disparity, it is necessary to redress the political imbalance. This is the true nature of the conflict in the Horn of Africa, whether it be fought in the name of nation, region, ethnicity, religion, or clan.

The limitations of ethnicity as an analytical concept, and the true nature of the conflict become obvious in the cases treated below. The conflict in Eritrea and Southern Sudan involved not ethnic groups, but regions inhabited by several ethnic groups which made common cause in the struggle for power. In the case of Somalia, ethnicity is not a factor, since all Somali belong to the same ethnic group. In that contest, where clans appeared as the adversaries, the struggle was about state power and the resources it commands. By contrast, in Djibouti the adversaries were ethnic groups, yet the enclave had no experience of cultural hegemony or suppression of ethnic identities. However, in recent years it experienced the gradual concentration of state power in the hands of the Issa Somali, and the parallel political emasculation of the Afar. This was the cause of the conflict that erupted there in the early 1990s.

5

Regional Movements

5.1 Profile

The oldest and most serious conflicts in the Horn involved movements whose constituencies are regional rather than ethnic. The conflict in Eritrea began in the early 1960s, and continued with escalating intensity for nearly three decades. It was a major contributing factor in the demise of two Ethiopian regimes during that period: the collapse of the imperial regime in 1974, and the overthrow of its military successor in 1991. The demise of the latter brought the conflict to a halt, when Eritrea became independent. The conflict in Southern Sudan is just as old, albeit interrupted by a decade of peace during the 1970s. It also has been a major destabilizing factor in the tortuous political history of Sudan, and the principal reason for repeated interventions by the military in the affairs of the Sudanese state. The conflict in that region shows no promise of ending soon.

Neither Eritrea nor Southern Sudan has a deeply rooted historical identity that can explain the rise of a militant political movement identified with it. Both are recent and typically haphazard creations of European colonialism, neither having had prior existence as an entity of any sort. Eritrea was put together by Italian colonialism in the hasty manner characteristic of colonial creations throughout Africa. It comprises low and high lands, and includes pastoralists and peasants, Muslims and Christians, within its borders. Southern Sudan is a vast area inhabited by a variety of ethnic groups that were first brought under one rule by British colonialism, which also attached their region administratively to Northern Sudan.

Neither region is ethnically or culturally homogeneous. Eritrea is inhabited by a melange of ethnic groups, nine of which were officially recognized by the nationalist movement itself. The largest is of Abyssinian stock and represents the northernmost extension of its Tigray branch, kin to the people who live in Tigray province of Ethiopia, from whom they were separated when the colonial border was drawn along the Mareb River. Christian and Tigray-speaking, these highlanders till the land in a region plagued by soil erosion, drought and locusts. The flanks of the Eritrean plateau and the surrounding lowlands are inhabited by several groups of pastoralists, all of them Muslim. Among the largest are the Beni Amer in the west, a branch of the Beja family whose main base is Sudan's Eastern Province; the Afar in the southeast, part of a group whose other branches are found in Ethiopia and Djibouti; and the Saho, who inhabit the hinterland of Massawa and Arkiko.

Historically these two ports handled Ethiopia's trade with the outside world, and were home to Muslim merchant communities with close ties to the Arab world across the Red Sea. The ports and the lowlands came under Muslim control when the rising tide of Islam flooded the region more than a millennium ago, surrounding and isolating Abyssinian Christianity in its mountainous stronghold. Throughout this period, the Eritrean highlands were the embattled northern frontier of Christian Abyssinia, ruled by local vassals of the Abyssinian crown, but seldom under its direct control. Perennial conflict in this region found its ideological expression in religion, and has poisoned relations between followers of these faiths to this day.

The region below the 10th parallel North, known as Southern Sudan, occupies about one-third of the Sudanese Nile basin. It is home to many ethnic groups, the majority of which are grouped together by ethnographers under the name Nilotics, from the river that dominates their existence. The Dinka, Nuer, Shilluk and Anuak are the largest groups belonging to the same linguistic family and sharing basic cultural features. Pastoralism is the traditional Nilotic vocation, though some groups, like the Shilluk and Anuak, turned to cultivation some time in the past. The pastoralists combine transhumant livestock herding with the cultivation of grains.

Neither pastoralism nor cultivation produced a surplus, and Nilotic material culture remained primitive. Trade was not a part of the local economy, nor was there any sign of urbanization in this region before the colonial period. Nilotic society is segmented in typical pastoralist fashion, and their political organization reflected this pattern. The state of the Nuer, Evans-Prichard (1940) noted, might be described as 'ordered anarchy'. Far greater cohesion was exhibited by the Azande, the major non-Nilotic group inhabiting the southwestern rim of the Nile basin. An agricultural society, politically centralized under a monarch, the Azande had imposed their rule in what is now Equatoria Province before the colonial intrusion.

5.2 Colonial Origins

As elsewhere in Africa, colonial rule in Eritrea and Southern Sudan promoted a degree of regional integration through the establishment of institutions and processes linking the various districts and ethnic groups within each region, and provided them with a common vested interest in the existence of these institutions and processes. This trend was far more advanced in Eritrea than in Southern Sudan. Initially, Italian colonialism nourished grand illusions of turning Eritrea into an African Eldorado, where peasants from the impoverished Italian south could be transplanted. Italy invested heavily in infrastructure development, and provided Eritrea with an excellent road network, a railway system and a second port at Asab. In the 1930s, when Eritrea served as a base for the invasion of Ethiopia, the colony experienced spectacular economic and urban growth, and many thousands of Eritreans served in the colonial army. The Italian occupation of Ethiopia

(1936–41) provided a further boost, which continued until the end of World War II under a British military administration.

By 1945, Eritrea had an urbanization rate of 20%, by far the highest in the Horn of Africa, with nearly 30,000 industrial workers, and thousands of demobilized soldiers seeking a livelihood. It also had a trader class grown large on the transit trade with Ethiopia and, to a lesser extent, Sudan. Italian provision for education had been minimal. It did not extend beyond primary level, and Italian was used as the language of instruction, though some missionary schools used local languages and Arabic. The first secondary school was founded by the British administration in the 1940s, when the formation of an Eritrean intelligentsia began. Generally speaking, in the early postwar period, Eritrea had reached a higher level of development than Ethiopia. Asmara, its capital, was an attractive modern city of 100,000, linked by road and cable railway with Massawa, a major Red Sea port, and there were several more flourishing towns throughout the region.

By contrast, there was hardly any sign of development in Southern Sudan during the colonial period. Preoccupied with developing an agricultural export economy in Northern Sudan, where cotton became king, the British were generally content to maintain peace and collect taxes in the South. The pastoralist economy of the Nilotics was left unchanged; even veterinary services were not introduced here until the late 1940s. It was at the end of the same decade that the first and only agricultural project, the Zande Scheme, was introduced in Equatoria Province. The South remained outside the framework of the colonial economy, producing nothing new, exporting little more than it had during the pre-colonial period, but importing more, including food, to meet the requirements of the non-productive population that gathered around the colonial administrative centres. In contrast with the central riverine region in Northern Sudan, where a buoyant economy based on the production of agricultural exports underpinned a burgeoning urban sector, Southern Sudan remained primitive and impoverished. At independence, GDP per capita in the South was only half the average for the whole of Sudan. Urbanization received no boost either. In fact, it was positively discouraged for a time by a colonial officialdom worried about the effects of 'detribalization'. Juba, the region's principal town, had a population of 9,000 in 1956.

Though formally part of what was officially called the Anglo-Egyptian Condominium of Sudan, the South was separately administered and purposely insulated from the North. The main thrust of the so-called Southern Policy applied by the colonial regime until World War II was to exclude Arab and Islamic influence, ostensibly in order to protect the primitive Nilotics from exploitation and abuse by Northerners. In the 19th century, the South had been the hunting ground for Arab and European slave raiders and ivory traders from the North, an experience that left indelible memories in the South and confounds its relationship with the North to this day. The underlying purpose of the Southern Policy was to minimize Northern Sudanese influence in the South, in order to keep all options open for the eventual disposition of that region. One option was its incorporation into Uganda.

Accordingly, the South was made a closed area where permission was required for outsiders to reside or travel. The Equatoria Corps recruited only Southern soldiers, and even Northern Sudanese traders were squeezed out and replaced by Greek and Syrian merchants. Education in Southern Sudan was entrusted to missionary groups, who combined it with proselytization. As a result, the spread of Islam and the Arabic language was blocked. Christianity and English gained ground, but neither of these became dominant in a region where the bulk of the population, which numbered nearly 3 million in the 1940s, remained attached to traditional creeds and had no contact with the English language. Nevertheless, the incipient Southern Sudanese intelligentsia that made its appearance at the close of the colonial period had acquired a vested interest in these alien cultural elements, which became rivals of Islam and Arabic, the twin pillars of Northern Sudanese nationalism.

5.3 Decolonization

At the twilight of the colonial era, the future of both regions became a matter for bargaining among third parties. Ethiopia made a determined bid for Eritrea, and cultivated support among the people there and the Great Powers. Italy wanted to return to its former colony as a Trustee under the United Nations. Britain contemplated partitioning Eritrea in order to annex its western part to Sudan, and even Egypt considered putting in a bid for the port of Massawa. The United States was concerned with the future of Eritrea, because it was in the process of acquiring a military base in Asmara devoted to electronic intelligence, and was inclined to favour Ethiopia, whose government was anxious to become a US client. Hoping for a Communist electoral victory in Italy, the Soviet Union initially supported the Italian bid. When that hope was dashed in the 1948 elections, it supported Eritrean independence.

In the late 1940s, the people of Eritrea themselves divided on this issue, according to where each community perceived its interests lay. Understandably, the Muslims who made up half of the estimated one million population were, on the whole, loath to have any connection with Ethiopia, a country ruled by the traditional enemy, Abyssinia. Under the imperial regime in Ethiopia, their co-religionists were second-class citizens without access to state power or, in the traditional Abyssinian provinces, access to land. The expressed preference of Eritrean Muslims was for independence following a period of tutelage under the United Nations. However, the Muslim ranks were not solid. Certain pastoralist tribes that had lost traditional privileges due to the emancipation of their serfs in the 1940s supported Ethiopia, hoping the imperial regime would restore serfdom.

Understandably also, Christians generally favoured a link with Ethiopia, whose cause was enthusiastically promoted by the Christian clergy in Eritrea. The clergy was mobilized by the ecclesiastical hierarchy, which was hoping the imperial regime would restore land and other privileges the Church had lost during the Italian occupation. The enthusiasm of the Christians was moderated by the fact that they belong to the Tigray, the smaller branch of the Abyssinian family, with little power and few privileges in the state ruled by the Amhara, the senior Abyssinian branch. Under the imperial regime, the Tigray language was suppressed in Ethiopia along with all other tongues, in favour of Amharic. A small but vocal minority of Christians joined the Muslims in demanding independence. The late 1940s were a period of intense political strife in Eritrea, interspersed with violence. While the opposed camps fought for control of the colonial state, their rivalry found ideological expression in religion. In the words of a British colonial official, the people 'rallied under their rival religious banners and now stood divided against one another in opposing Moslem and Christian factions' (Trevaskis, 1960, p. 76).

As it became clear in retrospect, the Christian political leadership in Eritrea saw the Ethiopian connection as a guarantee of their own political predominance in Eritrea. The 'Unionists', as they were called, were certainly not aiming to turn Eritrea simply into another Ethiopian province ruled by relatives of Haile Selassie. As it turned out, this is what happened. The result was massive alienation among Christians, who rallied belatedly to the Eritrean nationalist movement, ensuring its eventual success.

First came a ten-year interlude (1952–62), when Eritrea was linked to Ethiopia in a federal scheme devised and imposed by the United Nations. The scheme provided for genuine Eritrean self-government and a democratic political system, complete with political parties, elections, free press and trade unions; all of which contrasted incongruously with the medieval imperial regime in Ethiopia. Muslims and Christians in Eritrea were guaranteed parity in public employment. The official languages were Arabic and Tigray, and each community was allowed to choose the language of instruction in the local school system. Although Arabic is the spoken language of only one small nomadic group in Eritrea, Muslims value it as the language of the Quran and of a high culture which is dominant in the region. All political groups in Eritrea accepted the compromise, and the Unionist Party formed the first regional government in 1952, with the support of a minor Muslim faction.

Ethiopia's imperial regime saw federation as the first step towards annexation. A flourishing democracy in a corner of the domain of an emperor who claimed to rule by divine right was the height of absurdity and could not be allowed to last, for it set a dangerous precedent for other regions of the imperial realm. Furthermore, the regime in Addis Ababa feared a future resurgence of the pro-independence movement in Eritrea. In a relentless campaign that involved intrigue, extortion, bribery and naked force, the imperial regime emasculated Eritrea's autonomy, demolished the democratic experiment, destroyed its own ally, the Unionist Party, and finally annexed Eritrea in 1962. The first resistance movement, the Eritrean Liberation Front (ELF), had been founded two years earlier.

The fate of Southern Sudan was decided without reference to the opinion of its people, who hardly had the chance to know what was being decided, let alone to influence the decision. Nationalism had an early start in Northern Sudan, where the Arab intelligentsia was inspired by the flowering of Egyptian nationalism in the interwar period. By the end of World War II, the nationalist movement – aptly named the General Congress of Graduates – had spawned two sizeable political parties competing to inherit power from the British. Britain abandoned its Southern Policy as a concession to the Northern Sudanese nationalists, hoping to wean them away from their attachment to Egypt. Southern Sudan was now integrated administratively with the North, with a view to merging this region into a unitary, centrally governed Sudanese state.

Northern Sudanese nationalism had no echo in the South. Given the isolation and retarded state of education in that region, a native intelligentsia had yet to make an appearance at the close of the colonial period. Neither of the two Northern nationalist parties established branches in the Southern region, nor sought to rally support there. In 1947, having decided to include the region in the Sudanese state, the colonial government gathered a small group of literate Southern administration employees and illiterate chiefs in Juba, to inform them of the decision, and asked them to approve it. Realizing this meant exchanging British for Arab rule, these Southerners voiced grave misgivings, arguing their region could not hold its own against the North in any field. They finally consented to take part in the legislative assembly, after being told that decisions affecting the South would be taken without Southern participation. The colonial government and the Northern Sudanese nationalists chose to interpret the consent of this miscellaneous gathering as an acceptance by the people of the region to be governed by Khartoum.

The 1952 revolution in Egypt that brought Gamal Abdel Nassir to power forced Britain to make precipitate concessions to Northern Sudanese nationalist demands for independence. Within a year, agreement was reached for Sudan to attain independence in 1956. There was no Southern Sudanese representation in the negotiations, since there was no political organization in that region at the time. Nevertheless, the handful of Southerners who had benefited from missionary education and were employed in lowly positions in the regional administrative service showed concern over the looming prospect of Arab domination. The initial demand from this group was for independence to be delayed, in order to give the South time to catch up with the North in education and other areas of development.

Southern fears of Arab domination seemed confirmed by the results of the 'Sudanization' of the civil service in 1954. The Sudanization Committee, in which Southerners were not represented, allocated jobs and made promotions on the basis of seniority, experience and qualifications. Southerners qualified for only eight subordinate posts out of eight hundred, and denounced this as the 'Arabization' of the state service. A newly formed Liberal Party in the South immediately issued a call for federation. This was to become the rallying cry of the South and a fixed item on the political agenda of Sudan to this day. Sudanization also affected the soldiers of the Equatoria Corps, now renamed the Southern Command. In the closing days of the colonial period, a dozen Southerners had been raised from the ranks to become second lieutenants. They and other rankers fully expected to fill some of the posts vacated by the departing British officers, but were bitterly disappointed when Northerners arrived to claim all of them.

To Southerners it seemed their country was passing from British to Arab rule with only token participation on their part. The prevailing atmosphere of suspicion and anxiety was exacerbated when Northern troops arrived in Juba, giving rise to rumours that Southern units were going to be disarmed. In August 1955, when a Southern unit stationed at Torit was ordered north to take part in the independence celebrations, its soldiers mutinied, and their example was followed by other garrisons in the South. An assault on Northern army officers and civil officials soon turned into a pogrom of Northerners generally, and more than 300 people lost their lives. Many of the mutineers fled to the bush, where they later formed the core of the *Anya-nya* guerrilla movement. Southern demands for federation were ignored by the Northern nationalist leadership, and Sudan attained independence in 1956 as a unitary state.

5.4 The Conflict

In both Eritrea and Southern Sudan, the evolution of the conflict followed the same escalating pattern. The regimes in Addis Ababa and Khartoum pursued the same two-pronged policy aimed at subduing regional resistance and integrating the regions into a unitary, centralized state. Force was the main instrument used in the effort to subdue the rebels. The military and security apparatus in Ethiopia and Sudan expanded rapidly in size, consuming a steadily increasing share of the state budget and gaining virtual autonomy from political control, until eventually the state itself was brought under military rule. The escalating use of force had ambivalent effects. On the one hand, it seemed able to contain the challenge, and managed to preserve both the state and the power monopoly of the ruling groups for a long time. On the other, state violence inflamed regional resistance and expanded its scope by driving people to join the ranks of the rebels. Military operations laid waste whole districts, drove people off the land, produced successive waves of refugees, disrupted the process of production and distribution of food, and caused massive starvation. Moreover, years of uninterrupted warfare saddled the state with enormous foreign debts, bringing it to the verge of bankruptcy.

The second prong of state policy was a campaign of 'national integration' intended to eliminate the cultural differences perceived to be the cause of dissent. Not surprisingly, what the ruling groups in Ethiopia and in Sudan perceived as the 'national' identity was their own ethnic identity writ large. Accordingly, Christianity and Amharic became the hallmark of Ethiopian nationalism, Islam and Arabic of the Sudanese nationalism, and national integration was premised on assimilation. Not surprisingly also, this premise was rejected by other ethnic groups, and when the regimes sought to impose it by force, it became a potent source of alienation and a great boost for dissident movements.

After the annexation of Eritrea, the imperial regime outlawed Tigray, the language of the Christian community, and imposed Amharic. This was resented even by the most ardent Eritrean supporters of Ethiopia, and alienated en masse the Eritrean student population who now had to master two foreign languages – English and Amharic – in order to complete secondary education. The students were the first section of the Christian community to join the rebel movement. The use of Arabic in education was squeezed out gradually in Eritrea by the simple technique of not training teachers for that language, despite the pleas of Muslims that it is essential for the practice of their religion.

By contrast, in Southern Sudan it was Christianity that came under attack. Missionary schools were taken over by the state in the first few years of independence; Friday replaced Sunday as the weekly day of rest in 1960; missionary activities were restricted beginning in 1961; and all foreign missionaries were expelled from Southern Sudan in 1964. A plan for education adopted in 1961 was premised on thorough Arabization. Northerners took over all school supervisory posts in the South, and the region's secondary schools were later removed to the North. Southerners in public service came under pressure to take Muslim names and to send their children to Quranic schools. These moves drove the budding Southern Sudanese intelligentsia to embrace the rebel movements known as the *Anya-nya* in the 1960s.

The resistance movements that emerged in Eritrea and Southern Sudan were truly regional in character. From the outset, their frame of reference was the regional entity that had been carved out by colonialism, and all groups living in it were perceived as having common interests threatened by those who controlled the state. This perceived commonalty was not based on ethnic affinity, religious affiliation, or linguistic unity, since the population of both regions is quite diversified along those lines. Moreover, ethnic and religious divisions were strongly manifested in the internal political life of these movements, and at certain times greatly weakened them. Nevertheless, ultimately such divisions were transcended in the effort to resist the absorption and subordination of the regions in states controlled by ethnic groups who regarded themselves as culturally superior, who denied others access to power, and monopolized the material and social resources controlled by the state. In the course of the long struggle, regional solidarity evolved into consciousness of a common identity that can, at least in the case of the Eritreans, be called national.

5.5 Eritrean Nationalism

The Eritrean movement had a distinctly sectarian origin. Founded in 1960, the Eritrean Liberation Front (ELF) was an exclusively Muslim affair for several years. This is understandable, since it was the Eritrean Muslim community that suffered first and most from the elimination of the federal arrangement and the region's incorporation into Ethiopia. While the organizers and leaders of the ELF were former politicians, merchants and young educated people, its guerrilla fighters initially came from the pastoralist tribes of western Eritrea, especially the Beni Amer. For several years, the movement restricted its activities to the predominantly Muslim lowlands of Eritrea. The ELF sought and found support among the Arab states of the region, especially Syria, Iraq and the oil-rich Gulf states, where Eritrea was vaguely perceived as a predominantly Muslim region and a potential recruit to the Arab camp. The leadership of the ELF encouraged such illusions by stressing Islam and using Arabic as its official language. After the fall of the first military regime (1958–64) in Sudan, the Eritrean rebels enjoyed unrestricted access to Sudanese territory, where they established clinics, repair shops, offices and warehouses, and used Port Sudan to bring in supplies.

It was not until the mid-1960s that Christians began to join the movement, and the ELF was able to expand into the highlands. Until then, the ELF guerrilla force was organized on a zonal basis, each zone comprising the territory of a major ethnic group which dominated the organization at that level. Zonal commanders were natives of the area, and recruitment was carried out within the zone. Consequently, there was a strong ethnic element in the structure of the movement. There was also keen rivalry among ethnic groups for control of the ELF, and when elections to leadership posts were held in the 1970s, they were regularly fought on an ethnic basis. The Christians initially came from two social groups – students and workers. The former were affected not only by the banning of their language and the heavy-handed rule of Amhara officials who came to govern Eritrea, but also by the political radicalism that infected the student population of Ethiopia as a whole, turning it into the imperial regime's most implacable opponent.

The workers were permanently alienated from the Haile Selassie regime by the brutal smashing of the nascent Eritrean trade union movement in the 1950s. The subsequent stagnation of the Eritrean economy also affected them adversely. Eritrea entered the federal phase with a depressed economy, lack of domestic capital and sizeable unemployment. Although Asmara was the second largest zone of industrial concentration in Ethiopia, with more than a third of the enterprises and a fifth of the labour force, Eritrea was fast falling behind the central region of Ethiopia, where the bulk of new capital investment was concentrated. The local economy was never able to absorb the labour supply available in an urban sector that was proportionately more than twice as large as Ethiopia's, and many thousands of Eritreans were compelled to migrate to Ethiopia, Sudan and the Arab Gulf states in search of employment. There, they were gradually brought into the network of ELF cells organized in these areas.

The growing influx of Christians in the late 1960s, especially radicalized students, created intolerable strains within the ELF, whose conservative leadership was wary of Ethiopian infiltration and feared losing control of the

organization it had founded. The radicals, among whom Christians predom-
inated, forced a split in the nationalist movement. In the early 1970s, they
created a second organization under their control, which eventually became
known as the Eritrean Peoples Liberation Front (EPLF). This new organization forged an amalgam of socialism and nationalism as its ideology,
proclaiming its mission to be a social as well as a national revolution, and its
goal Eritrean independence. Independence was not a negotiable issue. In
order to put it beyond discussion, the EPLF defined Eritrea's position as
colonial and claimed the right to self-determination exercised by all colonized
people.

Ideological commitment and clarity of vision in the EPLF were matched
with organizational efficiency, discipline and military effectiveness (Sherman,
1980). None of these had been conspicuous attributes of the ELF, which
steadily lost ground to its rival. Inevitably, the two clashed, and a civil war
was fought between them during the early 1970s. These were years of exu-
berant radicalism among the student population throughout Ethiopia, which
preceded the collapse of the *ancien régime* in 1974. The Ethiopian radicals
considered the Eritrean struggle a just one, and saw in the rebel movement an
ally against the Haile Selassie regime. Eritrean youth, particularly from the
Christian community, flocked to join the rival nationalist organizations.
Intensified Ethiopian repression in the region produced more recruits from all
sections of the population, and both the ELF and the EPLF gained in size.
They were still locked in internecine conflict when the imperial regime was
overthrown in 1974.

The youthful Ethiopian radicals who carried the political battle against the
imperial regime formally recognized the right of Eritreans to self-determination. Yet it was not they who inherited power, but a military faction
committed to the preservation, not only of the territorial integrity of the
Ethiopian state, but of its highly centralized and authoritarian character as
well. Consequently, no serious attempt was made to engage the Eritrean
nationalists in negotiations. Instead, an all-out effort was made to subdue
them by greater force. A savage battle fought inside Asmara at the end of 1974
produced horrors that turned the majority of Eritreans against Ethiopia.
From now on, the nationalist tide mounted, and within two years the ELF
and EPLF had seized control of the region, with the exception of its capital
and the two ports on the Red Sea.

Under the imperial regime, Ethiopia had built the largest military force in
sub-Saharan Africa, thanks to generous assistance from the United States.
This force had not proved able to eliminate rebellion, however, and under the
new regime Ethiopia's military establishment expanded tenfold. Its share of
the budget soon exceeded 50%, universal conscription was introduced in
1983, and huge quantities of weaponry were procured from the Soviet Union.
Successive massive campaigns were launched in Eritrea, where battles were
fought on a scale previously unknown in sub-Saharan Africa. The regime
had considerable initial success. In 1978, it was able to recover most of
Eritrea from the nationalists. The latter were forced to retreat to inaccessible

areas, from where they could not be dislodged despite determined Ethiopian efforts.

The destruction wrought in that region provided the nationalist movement with a steady stream of recruits, while weapons of increasing sophistication were routinely captured from the Ethiopian army. While repulsing the Ethiopian offensives, the rival Eritrean fronts also fought their own internecine struggle. This ended in 1981, when the EPLF, by now far the stronger of the two, defeated and drove its rival out of Eritrea. Subsequently, the ELF disintegrated into various squabbling factions and ceased to be a significant force. After several years on the defensive, the EPLF was able to go on the offensive in the mid-1980s, against a regime that had failed to gain legitimacy in a country stunned by a second famine visitation within a decade and threatened by economic paralysis. Forced conscription of soldiers and bloody purges of officers did little to improve the morale and efficiency of the Ethiopian army. The regime was further distracted by a rebellion in Tigray Province, where it was losing control of the area and the land routes to Eritrea. After a series of stunning defeats in the late 1980s, the Ethiopian army once more was left besieged in Asmara and a few other towns, with the rest of Eritrea under the control of the EPLF.

Despite the weakened condition of the regime, the EPLF made no attempt to capture these towns, fearing their destruction; a fate suffered by the port of Massawa when it was captured by the Eritreans. Instead, the EPLF concentrated on a diplomatic offensive abroad, designed to win support for its proposal to settle the issue through a referendum in which the people of Eritrea would exercise the right of self-determination to decide their own future. Shunned until that point by all Western and African governments, the Eritrean leaders now gained a hearing in many capitals. With the end of the Cold War, the United States appeared resigned to the inevitability of Eritrea's independence and anxious to play a mediating role. Lesser powers followed its lead, and after three decades of desperate struggle, the Eritrean cause began to receive grudging recognition. The end came when the Soviet Union ceased to supply military aid to Ethiopia. The final scene of the drama occurred in 1991, when the regime in Addis Ababa collapsed under the assault of the Tigray rebels and their allies. The Ethiopian army in Eritrea disintegrated, and the EPLF assumed complete control. Two years later, a referendum produced a nearly unanimous vote for independence, and Eritrea became formally independent the same year.

Three decades of warfare had a devastating impact on the region, whose population was estimated at 2.6 million by the 1984 census but was considered much larger by the EPLF. A large part of this population was widely dispersed outside Eritrea's borders. The western lowlands were depopulated, and the Beni Amer pastoralists had moved across the border into eastern Sudan. Hundreds of thousands of Eritreans had sought refuge in Sudan since the first Ethiopian pacification campaign in 1967. Even larger numbers had fled in the same direction to escape starvation in the 1980s. Educated and skilled Eritreans emigrated en masse to the Middle East, Europe and North

America. The local economy was shattered, the incomparable road and railway system left by the Italians was demolished, Asmara's growth was stunted, and smaller towns stagnated and became dilapidated. One positive result, from the viewpoint of the nationalists, was the adherence of all sections of Eritrean society, regardless of ethnic, religious, linguistic and other differences, to the nationalist cause. A very strong consciousness of common identity and solidarity had been forged in the struggle, so that now it became possible to see the emergence of an Eritrean nation.

5.6 Southern Sudanese Resistance

The resistance movement in Southern Sudan made its appearance in 1963, following five years of military rule. This was a period during which normal political activities were outlawed, and the military regime pursued a policy of 'national integration' by promoting Islam and Arabic, while restricting Christian religious activity. The Sudanese military and security forces had a free hand in dealing with suspected opponents of this policy, and were ruthless in their treatment of the Southern intelligentsia, particularly those who had been involved in political activity earlier. Educated Southerners and political activists fled to Uganda and Congo (Zaire), where they established the political arm of the Southern resistance movement, the Sudan African National Union (SANU). Others, many former soldiers, policemen and prison warders among them, went to the bush, where they joined survivors of the Torit mutiny to form the guerrilla organization known as the *Anya-nya*.

The two wings of the Southern movement were only tenuously linked. The political activists abroad were not involved in the creation of the *Anya-nya*, and had no influence in the guerrilla movement. They tried to solicit support abroad, particularly among Africans, portraying the conflict as racial, and relying heavily on the theme of Arab oppression of black Africans. They succeeded in enlisting strong support among Christian missionary bodies and church organizations in the West, which took up the Southern cause and raised a chorus of protest internationally. However, the Southern activists abroad gradually neutralized themselves through endless squabbling and factionalism, in which tribalism played a conspicuous role.

For some years, the guerrilla force consisted of several poorly armed bands, formed along ethnic lines and operating in their home districts, with little communication or coordination between them, and without common leadership (Wakoson, 1984). Even so, they posed a challenge to the Sudanese army which moved en masse to the South, where it congregated in small towns, venturing out on retaliatory expeditions that rarely engaged the rebels, but frequently brought destruction to communities suspected of harbouring them. The failure to end the conflict was the spark which ignited the popular upheaval in 1964 in the North that caused the regime's collapse. After a brief and futile attempt to engage Southerners in a dialogue, Northern politicians redoubled their efforts to subdue the rebellion. Defence expenditure more than tripled in four years, with the Soviet Union providing weaponry and training.

The *Anya-nya* made considerable progress during the second half of the 1960s. Initially, the rebellion was stronger among the sedentary communities of western Equatoria Province. Gradually it spread among the Dinka, the largest ethnic group in the South, in the provinces of Bahr el-Ghazal and Upper Nile, as well as among the Nuer, Shilluk and Anuak in the east. The guerrilla bands in the west obtained arms from the defeated Simba rebels in the Congo (Zaire), where they also established training camps. In the east, they found sanctuary inside Ethiopian territory with the consent of the imperial regime, which hoped thus to pressure Khartoum to close its borders to the Eritreans.

The guerrilla bands were gradually integrated into a unified military structure, whose organizational format was nevertheless based on ethnicity. The Southern force was divided into provincial units which, in turn, were subdivided into ethnic sections. A unified command structure was established in 1970, headed by Joseph Lagu, a former officer of the Sudanese army. The *Anya-nya* arsenal was considerably strengthened thanks to Israeli assistance, and its military effectiveness greatly improved by the end of the 1960s. Israeli assistance to the Southern Sudanese was routed through Ethiopia, where Israel was also engaged in assisting the regime against the Eritreans, and through Uganda, where Israel had large military and economic aid missions until the rupture with Idi Amin.

In contrast to the rebellion in Eritrea, the Southern Sudanese movement was innocent of ideology. No great effort was made to conceptualize its goals in the ideological context of nationalism, and the closest Southern propaganda came to abstraction was a weak attempt to play on the theme of Arab victimization of Africans. The motto of a short-lived publication produced abroad was 'Negritude and Progress'. The leadership of the guerrilla force did not formulate a specific programme, nor did it ever specify the goal for which it was fighting, other than to pledge to rid the South of Arab domination. Many voices spoke for the South, including several factions of political activists organized in Sudan and abroad, and they were seldom in agreement.

The attitude of Southerners towards the Sudanese state, as distinct from the Arab hegemony, remained quite ambivalent. At a conference held in Khartoum in 1965, various Southern factions presented proposals ranging from independence to federation, and were easily outmanoeuvred by the Northerners, who were unanimous in rejecting all such demands. The one common denominator of Southern political opinion was self-government for their region. Such ambivalence made the conflict seem less than a fully fledged struggle for national liberation, since the Southerners were mainly concerned with regional autonomy, and this is what they settled for in 1972. Nevertheless, given the fact that the goal of the movement was to gain control of state power in the Southern region, it was essentially the same as other contemporary dissident movements in the Horn of Africa.

The offer of regional autonomy brought the conflict in Southern Sudan to

a negotiated end in 1972. This is the only instance of peaceful conflict reso-
lution in the Horn, albeit a temporary one. The offer was made by the second
military regime (1969–85), headed by Mohammed Gaafar Nimeiry. It came at
a time when *Anya-nya* military effectiveness was at its peak, and the state's
presence in the South had eroded to the point where the insurgents were pro-
ceeding to fill the vacuum with an alternative administration. It was the
realization that a military solution was not a realistic option, and awareness
of the political danger a protracted and unwinnable struggle posed for his
regime, that brought Nimeiry to the negotiation table in Addis Ababa in
1972. On the other side, supporters of the rebels, including the religious
bodies abroad and the regime in Addis Ababa, urged them to negotiate
(Mohammed Omer Beshir, 1975).

The Addis Ababa Agreement of 1972 succeeded at the time because it
addressed not only the basic demands of the Southern movement, but also
the political forces generated by the conflict. After an initial attempt to bypass
the *Anya-nya*, the Northern regime in Khartoum negotiated directly with
their leadership. Another contributing factor was the willingness of this leadership to accept a minimal, yet meaningful, offer. The South did not win
equal status with the North in a federal system, but attained a significant
measure of self-government within a unitary state. The three Southern
provinces formed a single region with its own legislature and executive
authority to manage regional affairs. The language issue was resolved by
making English the 'principal' language in the South, while Arabic remained
the official language of the state. Freedom of religion, and regional control of
education, offered assurances of protection against enforced Arabization.

The Agreement satisfied the outstanding demands of the social groups in
the South that had led the movement. The autonomy of the Southern
Regional civil service guaranteed a monopoly of state posts to the Southern
intelligentsia, thus removing a major grievance. Additionally, they were guaranteed a share of civil posts in the central government. Most of the *Anya-nya*
fighters were inducted in the Sudanese army in separate units stationed in the
South. Others joined the regional police and prison services, while their leaders became army officers, and the top commander of the *Anya-nya*, Joseph
Lagu, became a general in the Sudanese army. Finally, it was promised that
rapid development in the South would be promoted by a special development
plan financed by the central government and foreign aid.

The Addis Ababa Agreement brought peace to Southern Sudan for a
decade. However, it brought no development. The Sudanese economy
declined steadily under the Nimeiry regime, and the central government was
unable to fulfil its financial obligations to the South. Only a fraction of the
promised funds in the special development budget were handed over by the
central government, hardly enough to promote meaningful development.
Apart from some prestige projects like a university and a television station in
Juba, little was accomplished. The resources of the regional government were
insufficient to cover more than a minor part of its budget, and it gradually fell
into arrears in meeting its payroll.

Two major development projects connected with the region became matters of great controversy. One of these was the Jonglei Canal (see Chapter 2). The decision to commence its construction, announced in 1974, was greeted with protest demonstrations by students in the South, and with great misgivings among many others in Sudan and abroad. Southerners resented the fact that a decision of such momentous importance for their region was taken without their participation, and with scant consideration for the interest of the region's inhabitants. Although various incidental benefits for the South were envisaged in the plan, and these won the support of politicians and traders in the region, the profound impact on the ecology and local economy had not been seriously considered. The Executive Organ for the Development of the Jonglei Area showed scant regard for what it called, in a statement to the People's Assembly, 'the interests of a subsistence economy of a backward and unsophisticated people . . . which can no longer today be maintained nor encouraged in the face of demands for more food'.

Another controversy involved the location of a plant to refine the petroleum found at Bentiu in Bahr el Ghazal Province in 1978 (see Chapter 2). The site is located in the Southern Region, and Southerners naturally hoped their region would share the benefits of the discovery. The regime in Khartoum had other notions, and even sought to disguise the fact that the site was in the South. The South demanded the refinery be located in the vicinity of Bentiu, to provide a pole of development and tax revenue for the region. Northern opinion was very much against this choice. The Nimeiry regime found itself in a quandary; after floating several alternative schemes, it ended up arguing it was more profitable to have the oil refined abroad rather than to build a refinery in Sudan at all.

The Addis Ababa Agreement allowed the state president considerable freedom to intervene in the affairs of the regional government, and Nimeiry intervened destructively in Southern affairs, unravelling the agreement stitched in Addis Ababa. He imposed his own candidates for regional office, repeatedly dissolved the regional government, ordered changes in the boundaries of the Southern provinces, and in 1983 divided the region itself into three separate units, in an obvious move to reduce Southern solidarity and political power. In doing this, he exploited ethnic and provincial rivalries which had mounted steadily within the region during the 1970s. A major political cleavage had emerged between the Dinka pastoralists, the largest ethnic group, accounting roughly for half the population of Southern Sudan, and the sedentary groups of Equatoria Province. Even though the Dinka themselves were fragmented along provincial and clan lines that ruled out common action, their numbers and high political profile in the region aroused fears and allegations of Dinka domination. As a result, the Equatorians led by Joseph Lagu, the former *Anya-nya* leader, supported Nimeiry's decision to divide the region.

The second round of the civil war had started a year earlier, when the Sudan Peoples Liberation Movement (SPLM) made its appearance. Its origins were not dissimilar to those of the *Anya-nya*. In fact, many of the actors were the same, since they came from the ranks of the *Anya-nya* fighters who had been integrated in the Sudanese army ten years earlier. There had been several mutinous incidents among them and many desertions during that time. The earliest occurred in 1975, and led to the formation of a movement called *Anya-nya* II, whose adherents were mostly members of the Nuer ethnic group, and whose goal seemed to be separation from Sudan. From bases in Ethiopia, they carried out raids in Southern Sudan. In a precautionary move in 1982, the regime decided to transfer to the North army units from the South, the bulk of which consisted of former *Anya-nya* members. This led to more mutinies and desertions of Southern soldiers and officers, who congregated across the border in Ethiopia and founded the SPLM in 1983.

The armed wing of the SPLM, the Sudanese Peoples Liberation Army (SPLA), made its presence quickly felt in the Southern provinces of Upper Nile and Bahr el Ghazal. Initially its ranks were filled mainly from the Dinka ethnic group. The Nimeiry regime sought to capitalize on this by trying to turn the Nuer-dominated *Anya-nya* II against the SPLM. The two actually clashed, but the SPLM finally defeated and absorbed the main *Anya-nya* II force. Remnants of *Anya-nya* II went over to the regime and were turned into a militia unit. Traditionally wary of Dinka domination, Equatoria Province was the last to come under SPLM influence. Within a year of its founding, the new movement was operating aggressively throughout the South, and had scored well-publicized successes by shutting down the oil-drilling operations at Bentiu and forcing a halt in the construction of the Jonglei Canal.

The SPLM's growth was ensured by Nimeiry's next desperate move in search of political support. As his political fortunes declined, along with the downward slide of the economy, the dictator courted the support of the Muslim Brotherhood, implacable opponents of Southern autonomy. At their behest, he flirted with theocracy, initiated constitutional changes towards the establishment of an Islamic republic, and imposed *Sharia* law on a bewildered population. Although it was not clear whether this was to apply in the South, the first victims of the draconian law were Southerners living in the North. This signified the final abrogation of the Addis Ababa Agreement, and it cost Nimeiry the support of most Southern Sudanese, as well as many Northerners. The SPLM now won broad support among Southerners and considerable sympathy in the North for carrying the fight against an increasingly oppressive regime in Khartoum.

In no small degree, the effectiveness of the SPLM was due to support it received from Ethiopia, whose government once again seized the opportunity to produce a *quid pro quo* for inducing Sudan to close its borders to the Eritreans. There had been several meetings over the years between the rulers of the two states, and as many agreements signed and promises made not to allow their respective territories to be used for hostile activities against each other, but nothing ever came of them. Ethiopia now allowed the SPLM to maintain bases inside its borders, and provided it with an arsenal as well as training for its fighters. Moreover, it provided radio broadcast facilities that enabled the SPLM to carry its message to the people of Sudan.

That message was puzzling to those who had assumed the SPLM was simply another Southern resistance movement bent on loosening Khartoum's grip on that region. On the contrary, the SPLM, whose leadership displayed a degree of political sophistication missing in the old *Anya-nya*, distanced itself from the latter's exclusive preoccupation with the South. In his first radio address to the people of Sudan, in March 1984, the SPLM leader, John Garang, a former *Anya-nya* member and later a Sudanese army officer, addressed them as 'Nationalists, Patriots, Comrades, Fellow Countrymen'. The SPLM, he declared, was not a Southern Sudanese movement, but a movement representing all regions and peoples of Sudan, particularly the marginalized and underprivileged groups like the Nuba in the centre, the Beja in the east, the Fur in the west, and the impoverished urban masses in the North, as well as the people of the South.

For the SPLM, he went on to say, the enemy was the political system that had ruled Sudan since independence, and the ruling classes that created and maintained it for their own benefit. It was this system that divided the people of Sudan into religious, regional and ethnic factions in order to weaken them. This system was responsible for the Southern problem and all the problems confronting all of Sudan. Unless the system was changed, none of these problems could be solved. Consequently, it was futile to struggle to resolve the problems of the South in isolation. The problems of Southern Sudan 'can only be solved within the context of a united Sudan under a socialist system that affords democratic and human rights to all nationalities and guarantees freedom to all religions and beliefs' (Garang, 1987, p. 23). The slogans of the SPLM were 'National Unity', 'Socialism', 'Autonomy where and when necessary', and 'Religious Freedom'.

The SPLM message was calculated to rally support outside Southern Sudan, and was sympathetically received among the social groups in the North – workers, professionals, students – who were readying themselves for the uprising that ousted Nimeiry in March 1985. However, even they fully expected the SPLM to lower its sights and negotiate a deal for the South when the moment came. Nimeiry was succeeded by a military junta comprising the top officers of his army, and they appointed a non-political civilian cabinet. The new regime froze but did not repeal the *Sharia* law. To the SPLM it offered the restoration of the Addis Ababa Agreement, that is, a return to the *status quo ante*, and to its leader, John Garang, the post of vice-president of Sudan.

The outright rejection of the offer came as a considerable surprise, especially in view of the reasons given for it. Firstly, the SPLM rejected the transitional military regime that replaced Nimeiry, because it was headed by the very same people who had defended his regime and waged the bloody war against the SPLM. Secondly, the Addis Ababa Agreement concerned only the South, and the SPLM argued it was not the representative of that region and was not going to negotiate solely on matters concerning the South. Thirdly, it believed the subject of negotiations ought to be a new political system for Sudan, and called for discussions with all the groups that had forced

Nimeiry's ouster. Meanwhile, it continued squeezing the Sudanese army into a few Southern towns, while it took control of the countryside in that region.

The return to civilian rule did not change the situation. The elections of 1986 marked a return to the political *status quo ante*, giving first place to *Umma*, a party with a religious base and the central pillar of the system the SPLM vowed to destroy. Also committed by its constitution to make Sudan an Islamic state, *Umma* could not bring itself to abolish the *Sharia* laws, the key precondition for negotiations set by the SPLM. Now fully in control of the state, the Northern political elite did not take seriously the SPLM's vision of a 'New Sudan', and could not comprehend why it was not satisfied with the offer to control the South.

The war continued, with a malignant innovation added to it. Militarily and financially hard pressed, the government began to arm certain ethnic groups to fight the SPLM. The Baqqara, an Arab pastoralist group and traditional supporters of *Umma* in Southern Kordofan, had perennially feuded with their Dinka neighbours over land and water. They were now armed by the government, presumably to defend their district against SPLM incursions. In fact, the Baqqara used the arms to raid deep into Dinka territory, taking large numbers of animals as loot. In an appalling return to the horrors of past centuries, Southern children were abducted to be kept or sold as slaves. Such ethnic 'militia' groups were organized also in the South to combat the SPLM.

By this time, the Sudanese economy was nearing bankruptcy. Basic services were faltering even in Khartoum, while famine threatened the entire country and was already taking a rising toll in the South. In the far west, the conflict in Chad spilled over into Darfur, complicated by a new confrontation over land between the Fur peasantry and Arab pastoralists (see Chapter 1, section 1.4). *Umma*'s political rivals in the North made overtures to the SPLM and advocated negotiations. Having lost hope of winning the war, the Sudanese army leadership itself demanded the government bring the war to a negotiated end, and the latter was forced to begin negotiating in earnest. In July 1989, just when negotiations with the SPLM seemed about to bear fruit, a third military coup took place, led by middle-rank officers with strong Muslim Brotherhood connections. Its apparent goal was to abort the negotiations and to continue the effort to subdue the SPLM by force.

After some months of equivocation, the ideological orientation of the new regime became clear. The National Islamic Front (NIF) emerged as the dominant political force in Sudan, and its leader, Hassan al-Turabi, became the ideological mentor of the military regime. *Sharia* law was reimposed and Sudan was declared an Islamic republic. The state apparatus was gradually infiltrated by the NIF, a large militia force manned by NIF militants was created to counter-balance the regular army whose officer core was drastically purged, and the country's economy was taken over by entrepreneurs associated with the NIF. The regime's energy and the country's resources were fully absorbed by the war, while the economy continued to deteriorate. The *National Economic Salvation Programme, 1990–1993* consisted of just eleven pages and contained hardly any figures (Sudan, 1990).

The war in the South continued, with government forces besieged in a few towns, including the region's largest urban centre, Juba. Supplying the towns became nearly impossible, and the huge civilian population that had sought refuge in them was starving. Neither the regime nor the SPLM were inclined to give priority to the relief effort mounted by international groups. The population of the war-torn region was devastated by famine, disease and violence. Millions of Southerners were forced to leave their homes. A large number found themselves in the North, living in festering slums, while others crowded into refugee camps in Ethiopia. According to one report over one-fifth of the population in the South perished during 1983–93 as a direct result of the conflict, and up to 80% of the region's population was displaced at some time (Burr, 1993).

The collapse of the military regime in Ethiopia in May 1991 proved a serious setback for the SPLM. The new regime in Addis Ababa, headed by the Tigray Peoples Liberation Front (TPLF), was indebted to Sudan, where it had found sanctuary during its long struggle against the military regime of Mengistu Haile Mariam. Moreover, the SPLM's close links with the Mengistu regime did not gain it favour among the new rulers of Ethiopia. As a result, the SPLM had to abandon its bases and facilities in Ethiopia in a hurry. Many thousands of Southern refugees also fled their camps in western Ethiopia and crossed into Southern Sudan in chaotic conditions, without a clear destination and without food, and frequently under fire by the Sudanese government forces. The SPLM lost its sanctuary in Ethiopia and the material support provided by Addis Ababa until then, as well as the broadcasting facilities. The areas under its control in the South were awash with displaced people wandering about the countryside, hounded by famine and violence. The SPLM was not only unable to take care of them, but it interfered, as did the Sudanese government forces, with the efforts of international agencies to provide food aid. 'Lifeline Sudan', a special operation mounted by a consortium of aid agencies to bring food into Southern Sudan from Kenya, had to run a gauntlet of harassment by the government forces and the rebel movement, which reduced its effectiveness considerably and wasted lives needlessly.

The strain created by the reversal of the SPLM's fortunes in 1991 exacerbated internal dissension and led to a split in the movement. In August of that year, a group that became known as the Nasir faction challenged both the leadership and the programme of the SPLM. This split appeared to be the result of several divisive factors. On one level, it was a challenge to the leadership of John Garang, who had led the SPLM since its inception. On another level, the split reflected traditional ethnic rivalries in Southern Sudan and the fear of Dinka domination. The leaders of the Nasir faction came from the Nuer ethnic group, from which they also seemed to draw the bulk of their support, while John Garang and many of his top lieutenants were Dinka. Ethnic animosities, especially between Dinka and Nuer, were prominent in the violent clashes between the two factions that followed the split.

Most significantly, the split revealed a fundamental difference concerning the future of the South and the goal of the movement. Garang's opponents argued it was futile to seek a fair deal for the South within the Sudanese state, in view of the determined refusal of the North to accept the South on a basis of equality. They saw little reason to pursue a struggle which had accomplished nothing but the destruction of their homeland. They argued the proper goal for the Southern movement ought to be separation from the North and independence for Southern Sudan. 'An independent South is the only solution,' proclaimed a news-sheet they issued (*Southern Sudan Vision*, no. 2, April 1992). Their argument was strengthened by the attitude of the traditional political parties of the North, now biding their time in feeble opposition to the military regime that had displaced them. Persistent attempts to form a common front with the SPLM on the basis of an agreed set of constitutional principles that would guide the Sudanese state in the future proved futile, because Northern politicians were unwilling to make the concessions demanded by the SPLM.* This proved that, as far as the South was concerned, there was little to choose between civilian and military regimes in the North. Garang had no choice left but to revise the goal of the movement, which now became self-determination for the South with the clear option of independence. It was a long road, littered with broken promises and dishonoured agreements, as one Southern leader noted, for Southern Sudan to reach this point (Alier, 1990).

---

*Such an agreement was reached in 1995, when the Northerners accepted the principle of self-determination for the South. By this time, the Sudanese regime itself had formally accepted the same principle, and was using it as the basis for negotiations with the Southern opponents of the SPLM.

6

Ethnic and Clan Movements

6.1 Introduction

Militant political movements whose constituency is explicitly ethnic, and whose goals include ethnic political recognition and autonomy, are found in Ethiopia. This chapter will examine movements representing the Somali, Tigray, Oromo and Afar people. All have played major roles in the many-sided conflict that has marked the recent history of the Horn. Similar in many respects, these movements also differ to an extent that amply illustrates the complex nature of ethnicity.

6.2 Somali Irredentism

The worst case of fragmentation resulting from the imperialist scramble in the Horn of Africa was inflicted on the Somali, who found themselves divided among no less than five states – Italian Somalia, French Djibouti, British Somaliland, British Kenya and Ethiopia. Decolonization did not heal the fragmentation of the Somali nation. Formed through the merger of only two of the five fragments – British Somaliland and Italian Somalia – the creation of the Somali Republic in 1960 did not satisfy the aspirations of Somali nationalism, whose ideological cornerstone was pan-Somali unity. This became a political imperative no Somali government could afford to ignore, and was enshrined in the flag of the Somali Republic in the shape of a five-pointed star, three of its points representing the unredeemed Somali homelands in the Horn. This terra irredenta comprised some 600,000 square kilometres, slightly less than the size (638,000 square kilometres) of the Somali Republic itself. It included the Ogaden and adjacent territories in southeastern Ethiopia – an area the nationalists called Western Somalia – as well as the Northeastern Province in Kenya (earlier known as the Northern Frontier District), and the Djibouti enclave, which in Somali nationalist parlance was called the Somali Coast.

The Somali state was born with a sense of grievous injury, and a ready cause for serious conflict with all its neighbours. To begin with, the Somali Republic refused to recognize the colonial borders that partitioned the Somali nation, while its neighbours, supported by a consensus among African states, upheld the geopolitical status quo. Its constitution granted citizenship to all ethnic Somali regardless of where they lived, and pledged to pursue the unification of all Somali. Despite the pacific professions included in the constitution concerning the pursuit of pan-Somali unity, it was obvious that only force could secure that goal. The search for weapons commenced immediately after the independence of the Somali Republic in 1960.

Ethiopia had already secured the patronage of the United States, and vigorously opposed the arming of its hostile neighbour. Other Western powers offered Somalia only limited military aid designed to preclude offensive operations. Dissatisfied, the Somali turned to the Soviet Union, and in 1963 signed an agreement which provided generous military aid. Alarmed, the Ethiopians demanded increased aid from the United States, and went on to build the largest military force in sub-Saharan Africa in the mid-1960s. In turn, Somalia's military expenditure increased from 14% of its budget in 1961 to 25.6% at the end of the decade (Stevens, 1976, p. 176). Thus, the arms race in the Horn began in the shadow of the Cold War.

The Ogaden was the focal point of Somali irredentism. Its large size (200,000 square kilometres) and population (estimated at between 500,000 and one million), the presence of the Haud pastures so integral to the pastoralist economy of many clans that inhabit the Somali Republic, and its geographic position like a wedge that threatens to split the Somali state into two, account for its primacy in Somali nationalist aspirations. Moreover, lying between Djibouti and northern Kenya, the Ogaden is the linchpin in the geopolitical pattern of the Horn. If it could be removed from Ethiopian control, the other two coveted regions could be recovered as well.

The Ogaden was conquered during the great Abyssinian expansion at the end of the last century, but was never really brought under effective Ethiopian control. It was the home region and bastion of Sayyid Mohammed Abdille Hassan, the Somali hero who led a protracted rebellion against the European and Ethiopian occupiers of the Somali homelands during the first two decades of this century. During the Italian occupation (1936–41), and subsequent British administration (1941–48), the Ogaden was administered jointly with the other Somali territories. It was returned to Ethiopia in 1948, although the Haud and a section adjoining the Djibouti–Addis Ababa railway line, known as the Reserved Area, were not returned until 1954.

The Ogaden seemed unaffected by Somali nationalist agitation in the early postwar period. The only town of consequence in the region was Jijiga, and here the Somali Youth League, the nationalist movement that emerged in southern Somalia under the British administration, opened a branch in the 1940s. It was closed in 1948, when the region reverted to Ethiopia, an occasion that provoked a Somali riot and bloodshed. The rest of the Ogaden was quite innocent of development. The Ethiopian state maintained a low profile there, and the nomads remained apparently unmoved by the nationalist aspirations of their Somali kinsmen across the border. One observer at the time thought the Ogaden were 'less influenced by Somali nationalism than their kinsmen' (Touval, 1963, p. 134). Another concluded what mattered most to them was not the national character of the state that claimed their territory, but the imperative of unhindered movement within it, and relief from taxation and other forms of administrative imposition (Lewis, 1958, p. 350).

Things began to change in the late 1950s, as the two neighbouring Somali colonies approached independence. Anticipating Somali irredentist claims, Ethiopia moved to strengthen its presence in the Ogaden, and also began to promote national integration. The first meant the imposition of restrictions and obligations highly objectionable to the nomads. The second involved the familiar demand for assimilation. The first elementary school in the Ogaden was opened in 1957, with Amharic, a language quite unknown in the region, as the medium of instruction. Having allocated a modest sum for development, the Ethiopian government deemed it appropriate to resume taxation of the pastoralists, a practice that had lapsed since the mid-1930s. When a head tax was announced in 1963, a spontaneous uprising broke out. The stage was set now for the first clash between Ethiopia and Somalia.

The Ogaden rebels solicited help from Mogadisho, where they got some weapons and a lot of encouragement. Radio Mogadisho broadcast appeals to all Somali in the unredeemed territories to rise in revolt, and drove people everywhere to a high pitch of excitement through exhortations, such as the song which vowed: 'I shall not feel well until we go to war to unite the Somali'. Employing guerrilla tactics, the Ogaden rebels had some initial success against the Ethiopian army, which was getting its first taste of such warfare. Their goal was to clear the Ethiopians from the region in order to unite it with the Somali Republic. There was no attempt to define their cause ideologically, save in terms of religion: the Ethiopians were labelled 'kafirs' ('infidels').

After a few frustrating months trying to subdue the uprising, Ethiopia turned its forces against the Somali Republic itself. The Somali pact with the Soviet Union was concluded in November 1963, and it was considered possible that, once armed with Soviet weapons, the Somali army would intervene in the Ogaden. Ethiopia brought troops to the Somali border, shelled villages and bombed towns inside Somalia, and threatened an invasion. The Somali army massed on the borders, and there was heavy fighting during the first three months of 1964. Superior Ethiopian force compelled the Somali to back down. After peace was negotiated in Khartoum, Mogadisho instructed the Ogaden to halt the rebellion in their homeland. Thus ended the first round.

Just as the rebellion in the Ogaden was reaching its climax, another uprising inspired by Somali irredentism erupted in Kenya's Northern Frontier District, on the eve of the colony's independence. That desiccated, remote region is home to a Somali pastoralist population that numbered about one-quarter of a million in the 1960s. As if distance, barrenness, and lack of transport and communication links were not enough to ensure isolation and neglect, the British colonial administration imposed additional barriers between the Northern Frontier District and the rest of Kenya. It declared the former a closed area to outsiders, and imposed a quarantine on the marketing of its livestock, thus inhibiting movement and trade between its inhabitants and the population of highland Kenya. No sign of development appeared in the District during the colonial period; the first elementary school there was not founded until 1946.

Not surprisingly, the inhabitants of the District were largely unaware of political developments in highland Kenya, and Kenyan nationalism found no echo among them. Neither of the two main nationalist parties – the Kenyan African National Union and the Kenyan African Democratic Union – had any presence in the District, or showed any interest in mobilizing support there. By contrast, the main Somali nationalist party, the Somali Youth League, established branches in the District and mobilized support for pan-Somali unity, until it was banned in 1948: the same year it was banned in the Ogaden.

In the approach to Kenya's independence, the Somali of the Northern Frontier District organized themselves to agitate for union with the Somali Republic. A commission of inquiry appointed by the British government in 1962 determined that the majority of the region's inhabitants favoured union with Somalia. Although it had been stipulated beforehand that the commission's findings would determine the future of the District, Kenyan nationalist pressure forced the British to renege, and the District remained within Kenya when it became independent at the end of 1963. The Somali-inhabited part of it became the Northeastern Province.

The uprising that was dubbed the Shifta ('bandit') war began a month before the celebrations of Kenya's independence in mid-December 1963. Bands of insurgents, armed by the Somali Republic, began attacking police stations in the District. Kenya's military establishment at the time consisted of some 3,000 men, hardly enough to meet the challenge posed by even a handful of armed dissidents in the remote and vast northern region. As a result, Kenya was forced to rely on British military assistance, and committed itself to a defence agreement (1964) which granted Britain military bases in its former colony. In return, Britain undertook to build up Kenya's military strength. Kenya also signed a defence treaty (1964) with Ethiopia aimed at their common foe, the Somali Republic. Britain committed air, communication and engineering units to Kenya, effectively guaranteeing the integrity of the new state. Thanks to this guarantee, Kenya was not compelled to embark on a programme of rapid militarization, a course of action rendered politically inadvisable by the army mutiny in 1964. As a result, Kenya spent less on defence than any of its neighbours in the Horn.

The uprising lasted four years. It wrought havoc in the region, and accelerated the decline of pastoralism. One indication of this was the negative population trend. The 1969 census recorded a drop of about 9% since 1962 for the Somali population, and 15% for the Borana Oromo, who had sided with the Somali in the conflict. The loss of livestock was far greater. Kenyan soldiers were disposed to regard all pastoralists as dissidents, and took a heavy toll by machine-gunning, looting and confiscating herds. Further losses were incurred due to the policy of forced settlement in strategic hamlets. Congestion of animals in limited spaces resulted in ecological disaster, and disease wiped out the herds. Thousands of destitute pastoralists became wards of relief agencies, while others turned to cultivation, or migrated to the urban centres of Kenya in search of a livelihood. By 1967, the insurgents had been defeated, and the government of the Somali Republic officially acknowledged the territorial integrity of Kenya.

A third struggle with Somali participation was fought in the 1960s in the Ethiopian provinces of Bale and southern Sidamo. Bale borders on the Ogaden and Somalia itself, and its southeastern portion is inhabited by Somali pastoralist clans, while the rest is inhabited by the Arsi, a branch of the Oromo people who, like the Somali, are Muslim. The Arsi were pastoralists until their lands and animals were expropriated by the Ethiopians after the conquest around the turn of the century, and many were forced to take up cultivation. There was no sign of development in this province, nor any urban centres. There were no roads, transport or communication facilities in the 1960s, not even electricity or a postal service. In the fields of health and education, Bale was the worst served of Ethiopia's 14 provinces.

Southern Sidamo marks the extreme point of Somali expansion westwards. Here they were pressing against another Oromo group, the Borana pastoralists. The Borana appealed for protection to the Ethiopian government, which was concerned with the inexorable Somali drift westwards, and now found a reason to order the Somali to clear out from Sidamo. Armed by the imperial regime, the Borana fell on the retreating Somali. In turn, Somali from southern Sidamo and Arsi from Bale journeyed to Mogadisho to solicit weapons, and fighting began in earnest in 1963. It soon spread among the Somali in eastern Bale, who were also agitated by the rebellion in progress at that time in the Ogaden, across the Shebeli River.

Somali and Oromo fought together against the Ethiopian forces in a struggle that outlasted the parallel rebellions in the Ogaden and the Northern Frontier District of Kenya. Arms and supplies came from the Somali Republic and encouragement from Radio Mogadisho. The rebels had no ideology other than their religion, and solicited support on the basis of a holy war against the Ethiopians. Nor did they formulate clear-cut objectives for their struggle, other than the removal of Ethiopian control from their region. Presumably, the Somali shared the irredentist aspirations of their kinsmen across the border, who included this region within the boundaries of Western Somalia. If the Arsi Oromo had any objections to this, they were not known. The rebellion lasted until 1969, when Mogadisho ceased to provide material support.

The conflict in the Ogaden and Bale flared up again in the mid-1970s, and this time it led to a full-fledged war between Ethiopia and Somalia. Following the end of the first round in 1964, the leadership of the insurgents in the Ogaden, Bale and southern Sidamo had taken refuge in Mogadisho, where the Somali government provided housing, jobs and subsidies. They formed loosely organized groups and waited for an opportunity to fight again. This opportunity came in the mid-1970s, when Ethiopia was in the midst of a political upheaval following the overthrow of the imperial regime in 1974. The military regime that succeeded it was besieged by enemies, and its own ranks were badly divided. Moreover, it was in the process of losing Eritrea to the nationalist rebels there, while its radicalism and Marxist rhetoric had alienated the United States, thus losing Ethiopia its patron and arms supplier.

Ethiopia's political and military weakness was an irresistible temptation for the Somali, and the military regime in Mogadisho came under great pressure not to miss the opportunity to recover Western Somalia. It responded by arming and training bands of refugees from the Ogaden, Bale and southern Sidamo, and sent them into Ethiopia to commence guerrilla operations in 1975. They were grouped into two separate organizations: the Western Somalia Liberation Front (WSLF) represented the Ogaden, and the Somali and Abo Liberation Front (SALF) represented the Somali and Arsi in Bale and southern Sidamo.

Although their membership and leadership consisted of natives of the contested regions, many of them veterans of the first round, the organizations themselves were creatures of the Somali regime and operated under its close supervision. The WSLF claimed to be fighting against Ethiopian colonialism and for self-determination of the Ogaden, although it was quite obvious that Somali irredentism provided the inspiration. As in the 1960s, the Somali and Arsi in SALF had no clear perception of the future. The territorial claims made by the SALF overlapped considerably with those made by the WSLF; a contradiction that did not seem to bother either of them or their sponsors in Mogadisho.

After the guerrilla units had probed the weak points of the Ethiopian defence, the Somali regime sent in its own military units in the summer of 1977, ill-disguised as guerrillas. When these failed to attain their objectives, a full-scale invasion of the Ogaden and Bale was launched by the Somali army. One reason for the timing of the invasion was the knowledge that Ethiopia had succeeded in replacing United States patronage with that of the Soviet Union. An agreement had been reached in December 1976, but Soviet weaponry would not arrive until autumn 1977. The Somali forces scored spectacular initial successes, forcing the Ethiopian army to retreat to Harar. This was a short-lived achievement, for only a few months later the Somali were routed and expelled by an Ethiopian counter-offensive, supplied and advised by the Soviet Union and reinforced with Cuban soldiers.

Victory helped the military regime in Addis Ababa to consolidate its position. It was able to eliminate the civilian opposition in the centre, and then launched an offensive in Eritrea to clear the nationalist rebels from the cities they had captured in that region. Defeat had the opposite effect on the fortunes of the military regime in Mogadisho, which had enjoyed a measure of public support until then. This support began to dissolve quickly after the debacle in the Ogaden. Disaffection was manifested within the ranks of the military in successive mutinies, followed by executions, imprisonment and the flight abroad of military personnel.

They were joined in exile by a steady stream of politicians, bureaucrats, intellectuals, businessmen, and others who fell foul of a regime that became increasingly oppressive and violent as its insecurity increased. Having earlier declared war on 'tribalism', Siad Barre was now forced to fall back on the support of his own clan, the Marehan, as well as the Ogaden and the Dulbahande clans with which he also had kinship ties, forging a triple alliance dubbed the MOD by his opponents. In return, as one Somali put it, he divided the country's wealth among them, with 'the lion's share going to the Marehan, the leopard's share to the Ogaden, and the hyena's share to the Dulbahande' (Said Samatar, 1983, p. 6).

6.3 Somali Clan Conflict

Inevitably, the opposition that emerged in Somalia in the early 1980s also mobilized on a clan basis. The first opposition organization, the Somali Salvation Democratic Front (SSDF), was dominated by the Mijertein, a clan in the South which had been politically prominent during the parliamentary period. By contrast, the Somali National Movement (SNM) represented mainly the Ishaq, the dominant clan family in the North, who had long been at odds with Mogadisho. Northern Somalia is a barren, thinly populated region, inhabited mainly by pastoralists. British colonial rule introduced few changes, and independence brought few benefits. Although Somalia is pre-dominantly a pastoralist country, its development policy was no different from that of most other countries in Africa. Capital and technological inno-vation were directed to the development of export crops, and to the urban sector for the establishment of import substitution industries. It was southern Somalia that benefited from these, especially the region of Mogadisho. The Ishaq were not in the mainstream of Somali politics. Early on they had their own nationalist organization and did not support the Somali Youth League. This meant they were left out of the ruling coalition formed by the Somali Youth League after independence. The first northerner to gain political prominence was Ibrahim Egal, who became Prime Minister in 1967, only to be overthrown by the military coup d'etat in 1969. Afterwards, relations between the Ishaq and the regime in Mogadisho went from bad to worse.

Both the SNM and the SSDF sought and received support from the mili-tary regime in Ethiopia. They were allowed to establish bases there, were provided with arms and training as well as broadcasting facilities, and they commenced guerrilla operations against the regime in Mogadisho. Following the current fashion in the Horn, and also to please their Ethiopian patrons, both organizations adopted the Marxist revolutionary rhetoric. Despite the obvious need for it, there was no cooperation of any kind between them, something that was to become the rule for the clan-based opposition move-ments that mushroomed in Somalia in the late 1980s.

As the Somali civil war unfolded, the Ogaden found themselves enmeshed in it, to the great detriment of their struggle against Ethiopia. Considered a pillar of the Siad Barre regime, they came under attack by the opposition, particularly the Ishaq-dominated SNM. The Ishaq and the Ogaden had been feuding for nearly a century over the prized Haud pastureland, which lies within Ethiopia but has always been used seasonally by the Ishaq. While the

Ogaden was under British control (1941–48), the Ishaq took advantage of British protection to extend their visits to the Haud to the discomfort of the Ogaden, who were inclined to regard the pastureland as their own preserve. When Ethiopia resumed control of the Haud in 1954, the Ogaden tried to reinforce their claims with Ethiopian support. An agreement secured by Britain to protect the rights of the Ishaq to the Haud soon broke down, and Ishaq–Ogaden clashes in the area became common.

During the short period in the mid-1970s when the WSLF was active in the Ogaden, the clansmen whom it armed took advantage of the situation to push the Ishaq out of the Haud. Then, in the 1980s, the Ishaq were able to repay their enemies with the same coin through the agency of the SNM. Assisted by SNM fighters, the Ethiopians managed to track down the remaining WSLF bands and to eliminate the guerrilla presence in the Ogaden. Thus, the Ishaq found themselves collaborating with Ethiopia, the traditional enemy of the Somali, to frustrate the prime goal of Somali nationalism, the recovery of the Ogaden.

Later, things were to get much worse for the Ishaq, the Ogaden and indeed most other Somali clans. While the SSDF lapsed into inactivity in the South, SNM incursions in the North brought harsh retaliation by the regime in Mogadisho, mainly against Ishaq civilians and property. By 1988, the beleaguered dictators in Addis Ababa and Mogadisho found common cause for detente. Siad Barre and Mengistu Haile Mariam met in Djibouti and agreed to stop supporting each other's enemies. The implied *quid pro quo* was Mogadisho's abandonment of the Ogaden, a poor recompense for this clan's support of Siad Barre.

Anticipating expulsion from Ethiopia as a result of the agreement, the SNM launched a daring attack against the main towns of northern Somalia. Prolonged, intensive fighting reduced Hargeisa and Berbera to rubble, and cost the lives of thousands. What shocked the Ishaq profoundly, and sealed their alienation from the Somali state, was the use of foreign mercenaries to fly the bombing raids that levelled the northern towns. Hundreds of thousands of Ishaq were forced to flee across the Ethiopian border into the Ogaden, an area that had seen many of its own people seek refuge in northern Somalia since the war in the mid-1970s.

By the end of the 1980s, the Somali state, the only one in black Africa to be blessed with a national identity, was dissolving into its clan components. As the collapse of the Siad Barre regime became a certainty, all clans began to raise armed units in order to compete for a share of state power. The Hawiye, the dominant clan in the Mogadisho region, finally put Siad Barre to flight in January 1991. Its leaders wasted no time in forming a government, without bothering to consult the other clans. Hawiye pretensions to speak for the Somali nation were quickly challenged by the rest, and clannish strife became endemic in southern Somalia.

Northern Somalia under SNM control remained relatively peaceful, though devastated by the fighting. Public opinion in that region favoured severing links with the South, and the power grab by the Hawiye in the South inflamed it more. The SNM leadership did not attempt to counter the public mood and, in May 1991, it declared the North an independent state under the name Somaliland. Thus, the merger that had brought the Somali state into being in 1960 came unstuck only three decades later. A little later, Ibrahim Egal, the prime minister who was ousted by Siad Barre in 1969, was chosen president of Somaliland.

For the Ogaden, Siad's deal with Mengistu at their expense was the last straw. Defectors from the moribund WSLF formed the Ogaden National Liberation Front (ONLF), and began to distance themselves from Somalia. Rejecting irredentism and the vision of pan-Somali unity, they strove to project a distinct Ogaden identity, declaring their goal to be an Ogaden independent of both Ethiopia and Somalia. In the political chaos prevailing in Somalia following the overthrow of Siad Barre, it was difficult to assess the strength of such sentiments. Nevertheless, the abandonment of Somali irredentist aspirations is not without precedent. After years of nationalist agitation inspired by the vision of bringing Djibouti into the Somali Republic, the Somali in that enclave voted overwhelmingly for Djibouti's independence in 1977. With the change of regime in Addis Ababa in 1991, the ONLF and WSLF accepted the offer of regional autonomy made by the new regime headed by the Tigray Peoples Liberation Front (TPLF) to all ethnic groups in Ethiopia.

Clan strife in the South led to the collapse of the Somali state. Mogadisho became the bone of contention between two Hawiye sub-clans, whose leaders vied for the leadership. A murderous struggle ensued, which left the capital in ruins, caused many deaths, blocked the port and airport, and raised waves of destitute refugees. A massive famine followed, because normal distribution channels were blocked, and the activities of foreign relief agencies were obstructed by the youthful gunmen who comprised the clan militia units. World concern mounted and led to a massive military intervention by the United States in 1992, under the auspices of the United Nations. The intervention was conceived in humanitarian terms, yet it employed military methods, and soon clashed violently with the Somali factions competing for control of the capital. Having failed to fulfil its humanitarian objectives, the United States withdrew its forces in 1995. It was followed by agencies of the United Nations, and the war-torn country was left more or less to its own devices.

The dissolution of the only state in black Africa which met the conventional criteria of national unity and qualified presumably as a nation-state baffled not only foreign observers but many Somali as well. The clan was obviously the key actor in this drama, and it was hard to understand why this unit of traditional society should emerge politically pre-eminent at this stage of Somali history, shattering the illusion of ethnic and national solidarity. The mythology of Somali state nationalism discounted the persisting significance of the clan, while extolling the importance of the state. However, in a society still predominantly pastoralist, it seems the clan retained its primacy. The state, on the other hand, with its roots in the urban sector, was of lesser significance for the pastoralists and failed to earn their highest loyalty (see Doornbos & Markakis, 1994). Pastoralist political alienation is the rule in the Horn, and nationalism apparently is no cure for it. With the disintegration of the post-colonial unitary, centralized, authoritarian state, the Somali clans were struggling for a share of power in a Somali state of the future, whose structure will obviously have to be quite different.

Since it was a struggle for state power, the clan conflict in Somalia was not essentially different from other conflicts in the region. Power is a means to an end, and some of the ends sought in this case were quite mundane. For example, the Hawiye sub-clans were feuding over the revenue accruing to a country's capital city, while other factions fought over income from ports, airports, roads etc. Perhaps the most important prize, and the least advertised, was land in the relatively fertile Juba–Shebeli valley. This region is the home of the sedentary Sab clans and other ethnic groups of Bantu origin. Land in this region became increasingly valuable as commercial agriculture with state promotion began to vie with livestock raising for predominance in the Somali economy. It became the target of the ruling groups in the country, which used the agency of the state to commence a process of dispossession of the local people that continues to this day. The process began in earnest with the land tenure reform enacted in 1975, which decreed a change from customary tenure to individual registration. The results are described in a study by Besteman (1994, p. 503):

> Local farmers [were] largely barred from registering their land because of the inequality of access to government and the incompatibility of the law with local land-use patterns. Land concentration in the hands of speculators generally uninterested in farming was increasing the tenure insecurity of local farmers and resulting in an increasing disparity in land ownership.

The largest share of land thus acquired went to the clans that made up the ruling coalition under Siad Barre. Rival clans got their chance after the regime's demise, when they moved in to dispossess the initial dispossessors, as well as the local people who were still holding land. Thus, as one rare report (African Rights, 1993b) on this matter points out, land grabbing was one of the factors involved both in the generation and continuation of clan conflict in Somalia.

6.4 Tigray Provincialism

The conflict that broke out in Tigray in the second half of the 1970s came as a considerable surprise, particularly in view of the dimensions it took. Surprise was due to the fact that Tigray is the very heartland of Abyssinia, and its past is part and parcel of Abyssinian history. Axum, the capital of the Axumite ancestors of the Abyssinians, lies in this province, with some remaining signs of ancient grandeur. The Tigray-speaking population extends northwards into the Eritrean highlands, the northern frontier of Abyssinia, where many wars were fought to keep invaders out. A century ago, a Tigray noble became the Emperor Yohannes of Abyssinia (1872–89), and waged war against the Egyptians and the Sudanese Mahdists. Other Tigray nobles married into the royal family and held important posts under Emperors Menelik and Haile Selassie.

Nevertheless, there are sharp differences within the Abyssinian family between the Tigray branch – much smaller in size, conservative and intro-verted – and the dominant Amhara branch who built and ruled the Ethiopian Empire, and mixed uninhibitedly with other ethnic groups. A basic difference is language. The Tigray language is related to Amharic, but quite distinct from it, and the two are not mutually intelligible. This difference has re-inforced a virile provincial consciousness and sense of identity that sets Tig-ray apart from other Abyssinian provinces. When the Tigray language was banned, it became the cause for deep resentment and the symbol of Amhara domination. The population of Tigray Province is not homogeneous. Nearly one-third are Muslims, and relations between the rival faiths in this province have never been easy (Firebrace & Smith, 1982, p. 48). Eastern Tigray is the home of Oromo agro-pastoralist groups, while the eastern flanks of the plateau and the lowlands are inhabited by Saho and Afar pastoralists.

Tigray was always a provincial contestant in the dynastic struggles of Abyssinia, and was ruled by its own nobility who were restless vassals of the Abyssinian crown. With rare exceptions, the head that wore the crown was Amhara, a fact that did not inspire steadfast loyalty among the Tigray. Historically, Tigray provincialism was manifested in strong opposition to the Amhara monarchy and the central government's claim to absolute authority. During the Italian occupation (1936–41), Tigray nobles did not prove immune to Italian blandishments, and after Haile Selassie's return in 1941 they were in disfavour. In 1943, the recently restored imperial regime sought to strengthen central control in Tigray by appointing Amhara governors and imposing taxes that had lapsed under the Italian occupation. This provoked a major uprising, which was put down after heavy fighting. Central control was imposed afterward, along with taxes. Worst of all was the banning of the Tigray language in a region where, as late as the mid-1970s, only 12.3% of the males claimed to speak Amharic and only 7.7% could read it (Hunting Technical Services, 1975).

Stark impoverishment underlay Tigray political alienation. With a popu-lation estimated in the 1984 census at 2.4 million, Tigray was the poorest Abyssinian province and probably the poorest in Ethiopia. Its land was dev-astated by soil erosion, and yields were half of what they were elsewhere in Ethiopia. Land fragmentation rendered peasant holdings minuscule. In the 1960s, close to one half of the holdings were less than half a hectare, and two-thirds were less than one hectare (see Markakis, 1974). Producing one crop annually, these were barely sufficient to maintain the peasantry at subsis-tence level in good years. Tigray had the highest incidence of drought in Ethiopia this century, and this phenomenon was invariably followed by famine. There were major famines in 1958–59, 1972–74, 1983–85, and minor ones in-between. Tigray peasants were accustomed to trekking far and wide during hard times in search of jobs and food. By outlawing hired labour, the 1975 land reform closed this outlet during the crisis of 1983–85, and forced the famine-stricken peasantry to seek refuge in Sudan en masse.

The people of Tigray, linking the misfortunes that befell their homeland to its political impotence, blamed the Amhara for their misery. Under the imperial regime, precious little investment for economic development went to the old Abyssinian provinces, and to Tigray none at all. There was not a single industrial establishment in the entire province, and an abattoir in the provincial capital, Makale, was the largest enterprise. Educated Tigray and those who managed to acquire some skills were compelled to seek employment outside their province, which had the highest rate of internal emigration within Ethiopia. They congregated in Addis Ababa, where also students from Tigray attended technical schools and the university. They frequented the same places, where the staple topic of conversation was the plight of their homeland. The horror of the 1972–74 famine, and the imperial regime's callous indifference to it, heightened their concern.

It was from within the Tigray student group at the university that the organization eventually to become known as the Tigray Peoples Liberation Front (TPLF) emerged. Its founders were a group of radicals active in the Ethiopian student movement, who shared the Marxist ideological convictions of their peers. Like them, they were profoundly frustrated by the military seizure of power in 1974, which shattered hopes for the transformation of the Ethiopian state. The majority of the Ethiopian student radicals, many Tigray among them, went on to challenge the military regime on what they considered a class basis, transcending ethnic and regional divisions. Their organization, the Ethiopian Peoples Revolutionary Party (EPRP), engaged the regime in urban guerrilla warfare, and was annihilated. Most of its members fled abroad. One group was taken by the TPLF under its wing and formed the Ethiopian Peoples Democratic Movement (EPDM), and yet another continued a guerrilla struggle in western Ethiopia.

A handful of Tigray university students had already formed a core group determined to fight on a 'national' basis, believing that under the circumstances this was the most effective way to struggle against oppression. Undoubtedly, they were influenced by the example of the Eritreans, from whom they sought and received invaluable assistance at the start. The TPLF was formed in 1975, and commenced guerrilla operations in Tigray Province the same year. Before it could establish itself in that region, it had to fight off several rivals. One of these was another Tigray organization called the Tigray Liberation Front, another was the EPRP, which had established an armed presence in Tigray, and yet another was the Ethiopian Democratic Union (EDU), an organization representing remnants of the *ancien régime*.

Preoccupied with the Eritreans, the Somali and the radical opposition at the centre, the military regime in Addis Ababa paid little attention to the incipient rebellion in Tigray. Reasoning that the insurgency in that province could not long survive the suppression of the revolution in Eritrea, it concentrated on achieving this objective for the next ten years. In Tigray, it was content to keep open the two main roads leading to Eritrea and to hold onto the towns along those routes. This allowed the TPLF to mobilize and train a guerrilla army composed mainly of peasant youth, and to rally the people of Tigray to its banner.

Like its mentor, the Eritrean Peoples Liberation Front (EPLF), the TPLF regarded itself as the instrument of social revolution in the Marxist mould, as well as the liberator of the people of Tigray from Amhara domination. Like the EPLF, its programme combined economic measures such as land reform and social objectives like female emancipation with the goal of self-determination for the people of Tigray, whom it considered one of the oppressed nationalities in Ethiopia. Initially, the TPLF seemed to lean towards secession and independence, but soon shifted its position to a middle ground, leaving open several options. It declared Tigray would remain within a democratic, multi-national, pluralistic and decentralized Ethiopian state, if such could be achieved. If not, Tigray would opt for independence. Later yet, setting aside the option of separation, the TPLF defined Ethiopia as a multi-ethnic state that belongs to all the groups living in it, and set its sights on reforming it.

Like its imperial predecessor, the military regime in Addis Ababa never seriously considered a political solution to the manifold conflict that threatened the Ethiopian state with disintegration, because such a solution required structural reforms, especially decentralization and regional autonomy. Like its counterpart in Khartoum, it was intent on even greater centralization of state power. Rhetoric aside, the political structures it created were designed to achieve just that. More than once it offered regional autonomy to the Eritreans, and the constitution it adopted in 1987 included an elaborate design for local self-rule. However, such offers were propaganda ploys, and not a single step was taken to implement them. Moreover, the insurgents rejected them because they had had nothing to do with their formulation.

After nearly ten years of stalemate in the armed conflict in Eritrea, the nationalist rebels there went on an offensive that wrested most of that region from the Ethiopian forces. The TPLF followed suit, and in the late 1980s won control of the entire province of Tigray. Unlike the Eritrean nationalists, however, the Tigray rebels did not remain content with that. The TPLF had nurtured a satellite movement called the Ethiopian Peoples Democratic Movement (EPDM) to represent an Ethiopian multi-national constituency, which presumably gave it the right to operate everywhere in that country. The core membership of the EPDM consisted of former EPRP members. The activities of the EPDM were focused in the province of Wolo, adjacent to Tigray. As the decade drew to a close, a combined TPLF–EPDM offensive rolled the regime's forces as far south as Shoa Province, threatening Addis Ababa itself. By now, the TPLF had shifted its position to a definite rejection of secession, and urged all other dissident movements in Ethiopia to do the same.

Massive foreign intervention, always a factor in the political affairs of the Horn, contributed greatly to the escalation and prolongation of the conflict.

The denouement of superpower rivalry in recent years contributed decisively to the collapse of the military regimes in Ethiopia and Somalia by depriving them of diplomatic, economic and military assistance. Both the United States and the Soviet Union, with their respective allies in Western and Eastern Europe, had been committed to the preservation of the unitary, centralized post-colonial state in the Horn, and materially supported the regimes that defended it. Neither superpower had any sympathy for the opposition movements. The end of the Cold War, with the deflation of Communism as a world movement and its devaluation as an ideology, dissipated US fear of revolutionary movements wedded to Marxism. Having long ignored the rebels in Ethiopia, the United States now became heavily involved in efforts to negotiate a peaceful solution to the conflict in that country. When these efforts failed, it actively encouraged the TPLF to take over power in Addis Ababa. The Soviet Union, on the other hand, sealed the fate of the Mengistu regime by withdrawing its political and material support, just as the United States abandoned the Siad Barre regime in Somalia to its fate.

The end came in May 1991. The regime's defences collapsed, its leader fled abroad, and the TPLF and EPDM forces, operating in a common front named the Ethiopian Peoples Revolutionary Democratic Front (EPRDF), entered Addis Ababa and assumed power. In its first step towards political legitimization, the new regime convened a national conference to which all existing ethnic organizations were invited. Ethnic groups that had never had a political organization were urged to form one in order to attend the conference. Ethnicity was to become the founding principle of political organization in Ethiopia. The conference, held in Addis Ababa in July 1991, was attended by representatives of some 20 political organizations, 15 of which claimed to represent ethnic groups. The conference adopted a charter which, inter alia, recognized the principle of self-determination for all ethnic groups in Ethiopia, and also recognized the de facto independence of Eritrea. Among the ethnic groups that participated in the conference and accepted the new order in Ethiopia were the Somali, who were represented by the veterans of the WSLF and SALF, as well as the ONLF. Their participation indicated a shift away from Somali irredentism. Many more ethnic political organizations were to appear in the months that followed. They took part in intense political activity during the first half of the 1990s, including elections of local, regional and central governments, and the adoption of a constitution for a decentralized federal state whose constituent units were ethnically defined. Regional and local units enjoyed wide powers of self-government, and were entitled to use their own official languages, while Amharic remained the official language of the state. Ethiopia was defined a multinational state, and a distinction was made between cultural (national) identity and citizenship. The empowerment of ethnicity was a bold attempt to resolve the conflict in that country. It remained to be seen whether it would succeed.

6.5 Oromo Nationalism

The Oromo are probably the largest ethnic group in the Horn. They are also the most widely dispersed. Historians are inclined to believe their dispersion was the result of a great expansion that occurred in the 16th century and took the bulk of the Oromo, who were then pastoralists, to central, eastern and southern Ethiopia, some large branches to northern Ethiopia, including Tigray Province, and smaller ones to northern Kenya. This view is challenged by Oromo intellectuals who believe the Oromo were present in this region much earlier (Mohammed Hassen, 1990). This is a disagreement with political overtones. The former view tallies with Abyssinian historical tradition, which sees the Oromo as recent intruders into Abyssinian domain, while politically sensitive Oromo maintain the Amhara invaded lands inhabited by Oromo since time immemorial.

   Whatever the truth may be, dispersion resulted in considerable differentiation among the various Oromo groups, as they adapted to different ecological conditions and mingled with other ethnic groups. The southernmost groups located in the arid zone retained the pastoralist vocation, while the rest took to cultivation on the highlands. Sedenterization forced modification of the traditional Oromo socio-economic organization, which was notably egalitarian and democratic, and some groups developed stratified societies and monarchical political systems. Many Oromo, including most of the pastoralists, espoused Islam.

   Other Oromo groups, particularly those that penetrated into the central and northern sections of the plateau, came into close contact with the Abyssinians. Many were converted to Christianity and some adopted Amharic as their tongue. In the 17th century, these Oromo groups figured prominently in the internal political affairs of Abyssinia, and for some time held the balance of power. In Shoa, in the centre of the plateau, the Oromo interacted intensively with the Amhara and to a large degree were assimilated. They assisted Emperor Menelik (1889–1913) in subjugating the rest of the Oromo and shared the spoils of conquest. Menelik's designated heir, Lij Yasu, was the son of an Oromo ruler.

   A massive land expropriation followed the conquest, when the Abyssinians took two-thirds of the conquered land for the state, leaving one-third for the subject population. The land claimed by the state was distributed to Menelik's officers and soldiers, Abyssinian nobles and high clerics, the royal family and its servants. The process of distributing land in the southern plateau continued until the end of the imperial era. Most of it went to Abyssinians, with a small share going to Oromo who had become largely assimilated into Abyssinian society.

   The result was massive dispossession of most Oromo peasants and other ethnic groups inhabiting the southern half of the plateau. The fact that the new landlords required labour and did not physically displace the indigenous cultivators masked the reality of land appropriation and softened the blow. The new landlords took a share of the produce and exacted a variety of fees and services from those who tilled the land. In fact, the latter became their tenants. The conjunction of ethnic and class divisions that resulted created a potentially explosive situation. Nevertheless, the imperial regime had little trouble ruling the Oromo peasantry, possibly because there had been no massive displacement and landlessness as a result of land expropriation. Moreover, the regime was able to enlist the cooperation of Oromo traditional chiefs to whom it granted landed possessions, tax exemptions and other perquisites.

During the long reign of Haile Selassie – he became regent in 1916, king in 1928, emperor in 1930, and was deposed in 1974 – integration and assimilation of the Oromo proceeded along two lines. One was the cooptation of Oromo notables into Abyssinian high society – Haile Selassie's own wife was of Oromo stock – and the absorption of educated Oromo into the ranks of the newly established bureaucracy and the military. The second was the involvement of Oromo in the modern sector of the economy which developed in the central region along the railway line, and in the coffee-producing region of the southwest. All these are areas of Oromo concentration, and the material welfare of their inhabitants was linked to the Ethiopian state.

Though some Oromo reached the highest levels in the Ethiopian state and society, they got there on their individual merits and Haile Selassie's favour, and were nothing more than imperial retainers. They did not represent any assertion of Oromo power in the imperial state. An incident in the mid-1960s revealed the limits of their position. It concerned a self-help association formed in 1963 with state approval, to promote development in Oromo areas of central Ethiopia. The association acquired unexpected popularity and momentum, and that aroused the suspicion of an imperial government wary of any sort of voluntary collective activity. The reaction was swift and brutal. When the association's leaders refused to cease their activities, they were accused of treason and sentenced to death in 1968; this sentence was in fact carried out on only one of them.

Despite its tragic ending, this initial attempt at Oromo collective action proved inspiring. While it lasted, the association aroused considerable interest among urban Oromo, particularly in Addis Ababa, and attracted a number of Oromo students studying at the university and secondary schools there. This was a time of political awakening and radicalism for a student generation that was to play the key role in undermining the imperial regime. For some Oromo students, their political awakening occurred in the drama of the self-help association, and it was infused with a rising ethnic political awareness. In the twilight years of the imperial regime, following the fashion in radical student circles, they organized groups to study Oromo history and culture, subjects that had been taboo until then in Ethiopia.

The Oromo Liberation Front was the last major actor to make its presence felt in the crowded political arena of the Horn, although its formation dated back to the mid-1970s. Initially, the collapse of the imperial regime had raised hopes of emancipation for the subordinate ethnic groups. The new military regime did not share the Abyssinian chauvinism of its predecessor, and its own membership included a number of Oromo, as well as members of other minority groups. Its proclamations condemned the evil of past practices and recognized the equality of all cultures and faiths in the country. Islam was given official status, and the printing and broadcasting of other languages was allowed. The regime went so far as to recognize the right to self determination, and promised to promote local self-government.

For some time, a small group of Oromo activists devoted themselves to organizing Oromo cultural activities and producing publications in their native language. However, like their peers from Tigray who were organizing the TPLF at that time, soon they realized the new regime regarded the national issue as a cultural problem, not a political one, and was proposing to resolve it by allowing freedom of cultural expression. Its actions in Eritrea showed it was not going to change the structure of the Ethiopian state to allow ethnic or regional autonomy. They concluded the Oromo had to struggle for national liberation, and formed the Oromo Liberation Front (OLF) in 1976. Following the lead of the Eritreans, they defined the position of the Oromo in Ethiopia as colonial, and called for an armed struggle to liberate 'Oromia', a state whose borders were generously drawn to include all of central and most of southern Ethiopia (Markakis, 1987).

The OLF's chances of mobilizing broad support among the Oromo peasantry were undercut by the radical land reform enacted by the military regime a year earlier. The reform abolished land ownership and gave the land to the cultivators in usufruct. Landlordship and landlessness were eliminated overnight, the Abyssinian landlords left the countryside, and the peasantry rejoiced and praised the regime. It was a timely class action that dissolved the explosive ethnic-class conjunction, removed a major material grievance, gave the peasants a vested interest in the state and undermined the ethnic appeal of the OLF. The following year came the Somali invasion, and the Oromo rallied to the defence of the Ethiopian state. Indeed, Oromo peasants were to form the backbone of the Ethiopian army during the bloodstained 16-year reign of the military regime.

For the next few years, the OLF struggled to establish an armed presence in southeastern Ethiopia, without notable success. It had no greater success abroad, failing to secure sanctuary or aid from a neighbouring state. Somalia found the OLF's territorial claims a source of annoyance, because they overlapped with those of the Western Somalia Liberation Front and the Somali and Abo Liberation Front. An approach by the OLF to the largely Oromo leadership of SALF on the basis of ethnic solidarity was rebuffed. The Arsi Oromo from Bale felt they had more in common with their Somali co-religionists than with other Oromo in Ethiopia.

Later, the OLF shifted its attention to western Ethiopia. It was allowed to open an office in Khartoum, from where it established links with the EPLF in Eritrea. With some help from that quarter, it managed to establish an armed presence inside the western Ethiopian border. In the late 1980s, as the regime in Ethiopia became totally distracted by its reverses in the north, the OLF was able to expand its operations in that area. Over the years, the regime had

squandered its political credit with the peasantry. With political and economic chaos descending on the countryside, the OLF's appeal to the Oromo
began to evoke greater response. The size of its army expanded quickly, and
weapons were seized in large quantities from the Ethiopian army.

Like the EPLF, the OLF sought to put the goal of independence beyond discussion by defining Oromia as a 'colony' of Ethiopia. This was to prove a
contentious issue, not so much with the regime in Addis Ababa, which paid
scant attention to the OLF, but with other movements. Oddly enough, it was the
EPLF which took public issue with those who wanted to separate from
Ethiopia. Arguing that only Eritrea had been colonized by Ethiopia and was
thus entitled to unconditional independence, it urged all other movements to set
their sights on overthrowing the regime in Addis Ababa and reforming the
Ethiopian state, so as to enable all ethnic groups to live in freedom and equality.

This message was also intended for the TPLF, at a time when its ultimate
intentions remained unclear. Once the TPLF had settled its course, it also
opposed the OLF's separatist stand, and relations between the two movements were cool. They worsened when the TPLF sponsored a group called
the Oromo Peoples Democratic Organization (OPDO), formed by Oromo
soldiers of the Ethiopian army who had been taken prisoner by the TPLF.
When the TPLF and EPDM launched their drive southwards and entered
Oromo-inhabited territory, the OLF denounced it as an attempt to establish
Tigray hegemony. Nevertheless, the OLF was obliged to accept the new order,
and took part in the 1991 national conference. This seemed to indicate a
change of policy on the part of the OLF, whose avowed goal until then had
been Oromo independence. Several other organizations participated in the
conference as representatives of the Oromo people. These included remnants
of the Somali and Abo Liberation Front, now renamed the Abo Oromo
Front, a Muslim Oromo group, and the TPLF-sponsored OPDO. Initially,
the OLF became a partner in the transitional government, but later found
itself at a disadvantage in the political manoeuvring that followed, and moved
into belligerent opposition to the EPRDF regime. Meanwhile, the Oromia
regional government under the OPDO made Oromo the official language of
the region and promoted an Oromo cultural renaissance.

6.6 Afar Ethnicity

Like their Somali neighbours, the Afar pastoralists in the past comprised a
cultural community without political unity and suffered a fate similar to that
of the Somali during the imperialist intrusion in the Horn at the end of the
last century. The northern portion of the Danakil, their parched and arid
homeland, was claimed by the Italians and was included in Eritrea. The hinterland of the southern portion and the Awash river valley was annexed by
Ethiopia, while the Gulf of Tajura became the French colony of Djibouti.
This proved to be a permanent partition. The combined number of Afar in
the three states is estimated at about one million (Bonfiglioli, 1992, p. 134).

The extreme aridity of the Danakil offered little scope for development. Cultivation with the help of irrigation was traditionally practised in the Awash river delta, and salt was mined in the north and traded with the Abyssinian highlanders, who also bought animals from the Afar. Imperialist interest in the Danakil centred on its long coast on the Red Sea, and it was here that the Italians built the port of Asab, while the French founded Djibouti. Djibouti became Ethiopia's entrepôt when it was linked to Addis Ababa by rail in 1917. Asab came under Ethiopian rule when Eritrea was annexed by Ethiopia; it was linked by road to Addis Ababa, and became the site of an oil refinery constructed by the Soviet Union. Ethiopia lost Asab when Eritrea became independent in 1993, but secured an agreement that made it a free port.

The Awash was the first freshwater resource to be exploited in Ethiopia. It was utilized for electric power generation, and for the spectacular development of a commercial cultivation sector whose main products were sugar and cotton. Cotton produced on a narrow belt along the lower Awash occupied about 50,000 hectares by the mid-1970s. It was produced mainly by foreign companies under concessions granted by the Ethiopian government. The impact of this development on the Afar pastoralists of the Awsa region, who had no recognized rights on the land, was profound. The riverine area was their most valuable pastureland, the refuge for people and animals during the dry season that can last from mid-September to mid-June. Traditionally, an elaborate system of rules governed access to grazing and water points in this area (Maknun, 1993, p. 54). Its loss was a serious blow to the viability of the pastoral production system, as was shown when drought hit in the early 1970s. Denied access to the riverine area, the Afar lost perhaps one-third of their people and most of their animals (Bondestam, 1974).

Not all the Afar suffered from the intrusion of commercial cultivation. A group of Afar notables, headed by the family of Ali Mira, the Sultan of Awsa, were able to share in the wealth produced by cotton. They succeeded in putting land under cultivation for their own benefit, using the labour of dependants and impoverished herders. Their success was due partly to the indulgence of the Ethiopian government, which was wary of alienating the Afar leaders for fear of the disruption they could cause to cotton cultivation. Such disruption did occur in the second half of the 1970s, after the cotton plantations were nationalized, along with all land in the 1975 land reform. The Afar Liberation Front (ALF) appeared overnight, under the leadership of the Ali Mira family, and violence swept the area for some time before the Ethiopian military regime was able to regain control. The ALF was rendered quite ineffective by a split in 1976, when a group of defectors left to form the Afar National Liberation Movement (ANLM), which found common ground with the regime in Addis Ababa and was granted a measure of control in the Danakil.

Across the border, in the French colony of Djibouti, the Afar were involved in a power struggle with the Somali inhabitants of the enclave, as the colony approached independence. Although numerically inferior to the Afar at that time, the Somali were dominant in the town of Djibouti, were integrated in the commercial life of the colony, had a higher level of education, and were represented in the structure of the colonial administration. However, they did not enjoy the favour of the French, due to their association with pan-Somali nationalism whose aim was to unite Djibouti with the Somali Republic. The Front de Libération de la Côte Somalie (FLCS) was sponsored by Mogadisho for that purpose, and engaged in sporadic terrorist activity in the late 1960s and early 1970s. The French deported some Somali, changed the colony's name from Côte Française des Somalis to Territoire Française des Afars et des Issas, and installed an administration under Ali Aref Bourhan, an Afar. In a referendum conducted in 1966, a majority chose to remain under French hegemony. During the reign of Ali Aref (1960–76), a large number of Afar settled into the town of Djibouti, and became involved in direct economic and political competition with the Somali.

By the mid-1970s, the relative prosperity of Djibouti had moderated the irredentist fervour of the Somali in that colony, particularly as it had become quite clear that not only France, but also Ethiopia, as well as the Afar, were determinedly opposed to Somali irredentism. Furthermore, the heavy-handed manipulation of FLCS by the Siad Barre regime produced a split in that organization, part of which now advocated independence for Djibouti. In 1977, a majority of the Somali in Djibouti voted along with the Afar for independence. With the threat of Somali irredentism fading, the French now switched their favours to the Somali, particularly the Issa, the dominant clan in Djibouti. Ali Aref gave way to Hassan Gouled Aptidon, an Issa, who became the first president of independent Djibouti.

For a time, a rough balance was maintained in a volatile ethnic setting by sharing political office, not only between Afar and Somali, but also among the various Somali and Afar clans. The office of prime minister was reserved for the Afar, and cabinet seats were evenly divided. Djibouti maintained an artificial prosperity based on the activities of the port, the purchasing power of a 3,500-strong French military presence and twice as many European, mainly French, civilian expatriates. Djibouti's population expanded prodigiously, partly due to natural growth (estimated at 3%), and partly due to an influx of people fleeing the violence and famine that swept Ethiopia, Eritrea and Somalia in the 1980s. The bulk of the influx were Somali, who soon outnumbered the Afar by a large margin. The world economic recession and the political and economic bankruptcy of Somalia and Ethiopia in the late 1980s affected Djibouti's economy adversely, causing massive unemployment, particularly among the youth in the port's teeming slums. The collapse of the Somali state, the secession of Somaliland, the independence of Eritrea, and the political triumph of ethnicity in Ethiopia – all were momentous events that had an unsettling political impact on Djibouti.

Hassan Gouled Aptidon's long reign inevitably strengthened the Somali, and specifically his own Issa clan's, hold on the state. Gradually, they came to control most of the important posts and to enjoy the lion's share of the spoils of power. While the cake was growing smaller, the Issa were claiming a larger share, and the Afar became increasingly frustrated and resentful. Afar who were working with Hassan Gouled were regarded as puppets by their ethnic kinsmen, many of whom became convinced the regime had a sinister, long-range goal. In the words of one Afar author (Ali Goubba, 1993. p. 159), 'depuis l'indépendance, Hassan Gouled ambitionne et s'échine à faire de Djibouti une nation Issa, une entité Issa'. Under Hassan Gouled, Djibouti followed a familiar political path to authoritarianism, with a single party and a permanent president in whose hands were gathered all the reins of power. This situation fomented dissent not only among the Afar, but also among other Somali clans in Djibouti like the Gadabursi and the Ishaq.

The end of the Cold War, and the changing political climate in Africa in the early 1990s, brought demands for political liberalization in the mini-state. Afar politicians were at the forefront of the campaign to break the Issa monopoly of state power. Hassan Gouled responded with a mixture of fraud and force, in the manner that was becoming political fashion in Africa at the time. The result was civil war. Several opposition groups coalesced in 1991 to form the Front pour la Restauration de l'Unité et de la Démocratie (FRUD), which commenced guerrilla operations in northern Djibouti aimed at toppling Hassan Gouled. FRUD's demands included a new citizenship law to weed out late immigrants from Somalia. The leadership of FRUD was composed of Afar politicians, and its first president was a former prime minister of Djibouti. Its fighting force was mainly Afar, and it operated in the Afar region. Afar from Eritrea and Ethiopia appeared to have become involved, and the regime in Djibouti claimed to be fighting an invasion from abroad. France seemed unwilling to commit itself to either side, as the mini-state drifted into civil war. The regime raised an army of 16,000 men, probably the highest soldier/inhabitant ratio in the world, and led the country speedily towards bankruptcy. Deprived of political and economic support from its patron, the regime was forced to negotiate. A split in FRUD made its task easier, and a fragile peace was negotiated in 1995, with one faction of FRUD accepting office in Hassan Gouled's government. Other rebel factions vowed to continue the struggle, and Djibouti remained officially in 'a state of war'.

PART THREE
ENVIRONMENTAL SECURITY

7
What Is Being Done

7.1 National Efforts

Introduction

The prolonged, endemic conflict in the Horn of Africa has put additional stress on the region's already fragile environment. Direct damage has been inflicted in a variety of ways. The ecology of some areas has been damaged by forced population concentration, with increased human and animal pressure on the land as the result, as well as by the clearing of vegetation for military purposes, the extermination of wildlife and the erosion of land routes under heavy mechanized traffic. In other areas, the ecology has been negatively affected by forced depopulation, the cessation of normal productive activity and the invasion of the land by unsuitable plants and pests.

Direct damage has been limited. More serious are the accumulated problems the conflict has stored for the future, by preventing for decades any serious efforts to address the grave environmental issues confronting the region. With the exception of Kenya, conflict derailed the process of planning for environmental protection at the very start. Ongoing projects were drastically reduced or discontinued, while plans for new projects were shelved. Given the lack of institutional support, or even state control over large areas, environmental laws and regulations became meaningless exercises in formalism. In the reigning political instability and lack of accountability, those who wielded power were often among the chief transgressors of such laws. Furthermore, the massive human problems which the conflict has created will continue to divert attention and drain funds away from environmental protection in the foreseeable future as well.

Concern for environmental security is not lacking in the Horn of Africa. Drought and famine in recent years helped to focus attention on many aspects of environmental degradation and raised awareness of the dangers that lay ahead. All states in the region started putting in place the legislative and institutional framework for environmental protection. Implementation proved a major problem, not only because of the conflict, but also because of the usual scarcity of resources, inadequate organization and coordination, and lack of continuity that hamper most areas of state activity in developing countries.

In the developed world, it was private initiative, expressed through non-governmental organizations, which took the lead in raising public awareness and prompted state action to protect the environment. With the exception of Kenya, this has yet to make an impact in the Horn of Africa. State efforts at environmental protection are often undermined by ordinary people who find themselves compelled by urgent need, as with those who clear forest land in order to grow food, and cut trees in order to cook it. Motivated by greed, certain groups openly violate environmental protection laws, as in the case of those involved in hunting and trading wildlife, or the so-called 'suitcase farmers' in Sudan who degrade the land through over-cultivation.

Kenya

Environmental security is best promoted when it coincides with obvious economic interests of the state or powerful groups in society. This has been the case in Kenya, a country which took the lead in this field in the Horn, spurred by the economic importance of the tourist industry.

Wildlife and the Indian Ocean beaches have become the main tourist attractions. Appropriately, tourism and wildlife were made the joint responsibility of one ministry by the Wildlife Conservation and Management Act (1976). Originally, the Wildlife Conservation and Management Department in the Ministry of Tourism and Wildlife had responsibility for national parks, game reserves and biosphere reserves. In an effort to integrate all aspects of wildlife management, including its economic potential, this department was replaced in 1989 by the Kenya Wildlife Services, an autonomous parastatal agency authorized to draw up to 60% of its funding from game-park gate receipts. Under the direction of Dr Richard Leakey, Kenya Wildlife Services adopted innovative methods advocated by wildlife conservation specialists, including the involvement of local communities in managing wildlife as a sustainable resource (Adam & McShane, 1992).

Kenya is a signatory to the African Convention on the Conservation of Nature and Natural Resources (1968), the Convention on International Trade in Endangered Species of Wild Fauna and Flora (1973) and the Protocol Concerning Protected Areas and Wild Fauna and Flora in the Eastern Africa Region (1985). Hunting in Kenya – the land of the white hunter – was banned in 1977, and trade in wildlife trophies and products was outlawed a year later. Worldwide demand for such products continued, however, and so did poaching. The infrastructure that had been set up for sports hunting was now adapted to the booming business of eco-tourism. Prior to independence in 1964, Kenya had four national parks and six game reserves. By 1993, it boasted 26 national parks and 26 national reserves, plus several animal sanctuaries and nature reserves, occupying 4.4 million hectares, equal to 7.5% of its total land area. Endangered species such as the rhino, elephant, leopard and cheetah came under special protection.

While protecting and enhancing one of the country's main economic resources, the game parks and reserves deprived many pastoralist groups in

Kenya of valuable land and water resources. It was originally promised that pastoralists would retain free access to areas where they traditionally herded their animals. This promise was not always kept; and where outright eviction has not applied, pressure was brought on the local people to change their mode of livelihood – to become sedentary, to engage in the tourist trade, or to move elsewhere. For example, the Samburu National Reserve on the banks of the Ewaso Nyiro River forced many Samburu pastoralists to move out of the area by blocking their access to the river. Tourists now outnumber the indigenous people within the reserve. The sharp contrast between the ostentatious affluence of the tourists and the naked poverty of the local people is fraught with the potential for conflict. Ivory poaching and attacks on tourists in the reserve have been one result, making police protection a requirement for visitors to the area. In the Masai Mara Game Reserve, the government sought to compensate the pastoralists for the disruption of their way of life by imposing an additional charge on tourists, for the benefit of the local council.

An additional 3% of Kenya's land is gazetted forest. This does not provide full protection, however, since more than one-quarter of the annual supply of woodfuel and a good portion of timber come from there. Kenya has made a concerted effort to head off complete denudation by promoting afforestation. The Forest Department in the Ministry of Environment and Natural Resources was responsible for establishing industrial plantations and for reafforesting clear-cut areas. In order to involve local participation, the Rural Afforestation Extension Services programme was launched in 1978, and numerous state agencies also became involved. The Kenya Renewable Energy Development Project managed six agro-forestry centres which produced seedlings and provided training. The Kenya Woodfuel Development Programme initiated woodfuel projects at district level. The Presidential Commission on Soil Conservation and Afforestation was charged with coordination and monitoring soil and water conservation and afforestation activities throughout the state apparatus. Finally, the Presidential Commission Tree Fund was established to raise the finance required by tree-planting projects.

A number of Kenyan nongovernmental organizations provided support for local efforts, while the National Council of Women of Kenya through the Green Belt Movement sponsored the formation of women's tree-planting groups. Churches and schools were also involved (see Karuti, 1993). Fast growing, non-indigenous species like eucalyptus and citrus trees were preferred, and it was government policy to use improved germplast. Research in this field was carried out by the Kenya Forestry Research Institute. Forestry was taught at Moi University and the Lodiani Forestry College.

Agro-forestry was often used in conjunction with soil and water conservation programmes.* During the colonial period, various soil conservation

*This and the following two paragraphs are taken from Michael Stahl's contribution on 'Regional Cooperation to Prevent Soil Erosion in the Horn of Africa', presented at the PRIO/UNEP seminar in Sigtuna, Sweden, September 1991.

measures were imposed on farmers, and failure to comply was punished. As a result, soil conservation generated opposition, and was attacked by the nationalists in the campaign for independence. During the first decade of independence, soil conservation was neglected. Kenya officially recognized soil erosion as a serious problem in its contribution to the 1972 Stockholm Conference on the Environment. In 1974, a soil conservation project was initiated, gradually expanding to cover all the 42 districts in the country by the late 1970s. The Kenyan National Soil Conservation Project came under the Soil and Water Conservation Branch of the Ministry of Agriculture. Its goals included conservation of all arable and grazing lands in settled areas, increased production on small farms, the introduction of agro-forestry, and the use of labour-intensive methods.

Extension was the programme's major component: it included staff and farmer training, dissemination of training material, and supervision of field activities. Initially, the project worked solely with engineering measures such as terraces, check dams and cut-off drains. Later it was complemented with biological measures, such as grass strips and agro-forestry. An innovation of this project was the *fanya juu* method for terrace construction, which consists of digging a drain across the slope and throwing the excavated soil uphill to form a ridge. Initially, the project worked with individual farmers. Since land adjudication and registration have proceeded far in the highlands, most farms are private property, and it was felt the land tenure system guaranteed security for small landowners to invest in land improvement. In recent years, the project has adopted a 'catchment approach', whereby farmers in local communities are encouraged to cooperate.

By 1992, close to one million small farms had been terraced and had adopted other conservation measures. However, the total number of farms in need of conservation was estimated at two million. Most of these farms were located in the highlands, and the conservation programmes were designed for hilly areas with high rainfall. Numerous small-scale, local-level projects were launched under the Ministry of Reclamation and Development of Arid, Semi-Arid and Waste Lands. These included water storing and harvesting (catching run-off), land terracing, hedge building, mulching, minimum tillage, and tree and grass planting. The Local Afforestation Programme aimed to promote soil conservation through tree planting. The state also provided subsidies for soil and water conservation works. In 1988, the Kenya Agricultural Research Institute initiated a programme for soil and water management, and several universities in Kenya offered degrees in agricultural engineering.

Kenya has five marine parks and reserves on the Indian Ocean coast. The Department of Fisheries within the Ministry of Environment and Natural Resources was made responsible for the management of the country's inland and marine fisheries. The mangrove forests and the lowland rain forests are protected habitats. Nevertheless, the country's coral reef fringe and some marine estuaries were threatened by sedimentation discharged at river mouths. For instance, sedimentation from the Sabaki River affected the area south of Malindi which lies within two designated marine parks – the Malindi and Watamu Marine National Parks. Marine pollution was controlled by legislation banning the discharge of oil from ships, regulating on-shore uses of sea water, and mining activities within marine parks. A National Anti-Marine Pollution Committee consisting of representatives from government, the tourist industry and nongovernmental organizations was charged with responding to oil-spill incidents within Kenya's territorial waters and the 200 km Exclusive Economic Zone. Kenya is a signatory to the Protocol Concerning Cooperation in Combating Marine Pollution in Cases of Emergency in the Eastern African Region (1985), the International Convention for the Prevention of Pollution by Ships (1973), and the Convention on the Prevention of Marine Pollution by Dumping Wastes and other Matters (1972).

Responsibility for general policy on environmental security lay with the National Environment Secretariat at the Ministry of Environment and Natural Resources. One of its tasks was to formulate the country's general policy for environmental management; another, to assess the general environmental impact of proposed state projects. Other agencies, such as the ministries of Water Development, Health, Agriculture, Industry, each had sections charged with environmental protection. Created in 1974 within the Office of the President, and moved to the Ministry of Environment and Natural Resources when it was founded in 1979, the National Environment Secretariat initially lacked the power to enforce its own directives. Instead, it had to rely on the relevant ministries and state agencies, which did not always give the environment the same priority. Later, the National Environment Management and Enhancement Act was passed to empower the Secretariat to lay down standards and enforce them by administrative order or court action.

Sudan

Laws relating to environmental security in Sudan stem from three sources: customary law, local legislation and Islamic law. Customary law, mainly concerned with land and water rules, was enforced at the local level by traditional authorities until the abolition of the Native Administration system in 1971. Subsequently, the lack of staff and funds in the Local Government system resulted in a legislative and administrative vacuum which facilitated the process of land denudation and degradation, particularly in the savannah region. This was because the state claimed formal ownership of unregistered land – 95% of all land in Sudan – ignoring customary rights, especially in pastoralist areas, and was able to allocate large parcels to commercial farmers for commercial exploitation. Islamic law provided general principles for land, tree and water use, and some of these were applied in the formulation of the Civil Transactions Act of 1984. Many other statutes applied to specific issues, but there was no comprehensive legislation relating to environmental security.

The institutional setting was also largely missing. In the wake of the famine in the mid-1980s, the Relief and Rehabilitation Commission (RRC) was set up in 1986. Its brief was to deal with national disasters, defined as 'anything which affects the environment of the human being'. The RRC was to 'strive to maintain the natural resources and restore the environmental order in the areas affected by drought and desertification in cooperation with the competent bodies'. The RRC established a Soil Conservation, Land Use, and Water Programme Administration, which undertook to prepare a natural resources development plan. In the early 1970s, concern with desertification led to the founding of a Desert Control and Rehabilitation Programme (DECARP) in the Ministry of Agriculture, which was later integrated into the RRC. A National Plan for Combating Drought Effects and Desertification was submitted in 1985, but there was no follow-up. Some projects were launched in the 1980s in the vulnerable provinces of Kordofan and Darfur, and in the refugee-congested areas of Eastern Province, mainly by nongovernmental organizations, to promote tree planting in an effort to stem the advance of the desert.

Forest management was introduced in Sudan during the colonial period, and the National Forest Administration had its origins in that era. There is no forested area left in Northern Sudan, and the little over 4,000 square kilometres (0.2% of the total area) of forest left in the country was found in the South, mainly in Equatoria Province. Of this area, less than 200 square kilometres had conservation status. The Imatong Mountains tropical forest, covering an area of over 1,000 square kilometres, was gazetted as a forest reserve in 1952, but lacked conservation status despite its special ecological importance. During the period of regional self-government in the 1970s in Southern Sudan, the reserve was exploited by the regional forestry department, the Talanga Tea Project and the Imatong Mountains Development Company. There was a Forestry Research Centre at Soba, and the trend was to replace indigenous species with imported ones such as eucalyptus and cypress.

Wildlife protection in Sudan also had its origins in the colonial period. The Wild Animals Ordnance of 1935 formed the basis of subsequent legislation integrated into the Wildlife Conservation and National Parks Act of 1986. There were more than fifteen national parks and game reserves in the Sudan, only two of which – Dinder and Radoam – were in the North. They were administered by the Administration of Wildlife Conservation Forces, which included the Wildlife Research Centre. An aerial census of wildlife was made for the entire Sudan in 1977, and subsequently other surveys were carried out in the Southern Region. Wildlife was exploited in direct fashion through hunting. For the rural poor, wildlife was a source of meat. For the rich, hunting was a sport and a source of trophies. Others traded in live animals and animal products such as ivory, rhino horn and skins. Hunting was officially regulated and required a licence and the payment of fees. The regional government in Southern Sudan during the period of self-government had a Ministry of Tourism and Wildlife, and hunting by tourists was a major source of income. In practice, hunting in Sudan was out of control, decimating the animal population in the national parks and game reserves. Sudan exported live animals and animal skins, both legally and illegally. During the period 1979–88, Sudan exported 1,400 tons of ivory – nearly five times as much as the other countries in the Horn combined (UNEP, 1993, p. 208). Due to the continuing conflict in Southern Sudan, it was in that region that wildlife suffered the most.

The 1975 Environmental Health Act in Sudan stipulated that no solid, liquid or gas matter actually or potentially harmful to man or animal shall be put into the sea. Legislation passed in 1975 regulated fishing, as well as providing some protection for coral reefs. A revised fisheries act provided enabling legislation to permit the establishment of marine parks and reserves. In 1983, the Sanganeb Atoll northeast of Port Sudan was nominated by Sudan as a World Heritage Site. The area was declared closed to commercial fishing earlier, and plans were made to declare it the country's first national maritime park. The Suakin archipelago south of Port Sudan and the islands to the north are other sites that have been recommended as marine reserves. A focal point for marine conservation was the Institute of Oceanography of the National Council for Research.

A National Committee for the Environment within the National Council for Research provided a forum for discussion of environmental issues. There were several nongovernmental bodies concerned with the environment in the Sudan. One of them, the Sudan Marine Conservation Committee, lobbied for the introduction of new laws and more effective implementation of existing ones. The Sudanese Environmental Conservation Society was founded in 1975, and the Sudanese Wildlife Society in 1982. Wildlife clubs were formed in Southern Sudan in the 1970s. The Institute of Environmental Studies at the University of Khartoum offered a Master's Degree in that subject and promoted research and publications.

Ethiopia

Ethiopia's prime environmental problem is the degradation of the northern highlands, the Abyssinian homeland. This is an age-old process whose influence may be discerned in the southward drift of the Abyssinians over the centuries. Undoubtedly, it was a factor in the spectacular expansion of the Abyssinian state in the 19th century, and in the land policy applied to the conquered regions. Ever since then, northerners have moved to the southern plateau in a steady stream, a spontaneous movement designed to ease population pressure on the exhausted land of the northern plateau. The south, with the highest and most reliable rate of precipitation, has retained its verdant cover and some forested areas. Also, it was less densely populated than the north. Actively assisted by the state, this population movement amounted to a major effort to cope with environmental damage, although it was not normally perceived in these terms.

The first massive influx of northerners into the south came with the conquest of that region in the late 19th century. A major part of the land there was expropriated and distributed to Abyssinians. The practice of giving state grants of land to northerners in the south continued until the demise of the imperial regime in 1974. Land grants went normally to members of the ruling class and their retainers, not to peasants. Planned resettlement for poverty-stricken peasants from the north began in the 1950s, although only about 20,000 families had been resettled by 1974. A Settlement Authority was estab-lished in 1976, and was absorbed into the Famine Relief and Rehabilitation Commission in 1979. Resettlement proceeded at a gradual pace; about 100,000 people had been resettled by 1983. In the middle of that decade, and in the midst of an apocalyptic famine, the military regime launched a massive resettlement campaign estimated to have moved about 600,000 people within two years (1984–85). The settlers came from the famine-stricken northern region, mainly from Wolo and Tigray Provinces. Some of them were settled in the western fringe of the country, the rest in the south. This was a crash pro-gramme, more in the nature of an evacuation than planned resettlement, and it caused much suffering and loss of life. Most resettlement sites were aban-doned and little survived of that costly scheme after the collapse of the military regime in 1991. The same is true of villagization, another hastily conceived, costly scheme launched at the same time, which forced millions of peasants to abandon their homesteads in order to congregate in newly estab-lished villages.

A national soil conservation programme got under way in the late 1970s.* It was organized by the Community Forestry and Soil Conservation Department in the Ministry of Agriculture. The Department coordinated conservation activities in peasant communities. These activities included bunding and terracing, constructing check dams in gullies, hillside closure and afforestation. Until the conflict put an end to its activities, the Department had 300 field staff working in 117 catchment areas. Following a detailed study of land degradation in the early 1980s (FAO, 1984), which concluded that one-quarter of the land might be unable to sustain crop pro-duction by the year 2010 if erosion proceeded unabated, the Ministry of Agriculture launched a programme of rehabilitation in the northern high-lands. A notable feature of the Ethiopian soil conservation programme in the 1980s was its conjunction with the food-for-work programme funded by external aid, and the involvement of the peasantry in mass campaigns. The peasants were mobilized by their associations, and were compensated through a food-for-work programme funded by the World Food Programme. Other international agencies and donor governments provided technical assistance, tools, nursery supplies and transport. Several Ethiopian non-governmental organizations also assisted, among them the Ethiopia Red Cross and the Ethiopian Evangelical Church Mekane Yesus. Although this was one of the largest programmes of its kind in the world, it affected only a

---

*This and the next paragraph are taken from Michael Stahl's contribution on 'Regional Cooperation to Prevent Soil Erosion in the Horn of Africa', presented at the PRIO/UNEP sem-inar in Sigtuna, Sweden, September 1991.

very small portion of the highlands. Because of the conflict, it excluded the two most degraded areas, Eritrea and Tigray.

The programme's physical achievements were indeed impressive: more than one million kilometres of soil and stone bunds were constructed, 80,000 hectares of hillside were closed in, and 300,000 hectares were afforested. All the same, this amounted to less than one-tenth of the total area in need of conservation. An evaluation of the programme revealed various shortcomings, chief among them being the fact that terracing reduced the area available for cultivation – no mean consideration for peasants whose holdings were already minute. The closing of hillsides transferred grazing pressure onto nearby hills, with predictable results, and the survival rate of seedlings was around 40%. Significantly, the evaluation concluded the rehabilitation programme was not sustainable without food-for-work subsidies because the peasants would not continue to participate without them. The peasantry was alienated by a coercive regime; in addition to the insecurity imbued by the fact that the 1975 land reform gave the peasants usufruct rights over the land but not ownership, and the military regime had treated peasant land rights in a cavalier fashion (see Hultin, 1988; Stahl, 1988).

The example of Kenya prompted Ethiopia to commence planning the development of a tourist industry based on wildlife. It requested help from UNESCO to formulate a wildlife conservation policy, resulting in the establishment of a Wildlife Conservation Department and the designation of national parks. One was in the Semien Mountains in northern Ethiopia, the exclusive habitat of the Walia ibex, another was in the Awash valley, and a third in the Lower Omo valley, Ethiopia's most isolated and primitive region. The process of setting up the basic administrative structure for the national parks had only begun when it was overtaken by the conflict that engulfed the country.

The military regime that ruled Ethiopia during 1974–91 paid formal homage to the cause of environmental security. Official documents and reports were replete with promising statements, and the *Ten Year Plan, 1984/5–1994/5* included environmental protection and management among its primary objectives. The initial report of a *National Conservation Strategy* was prepared in 1990, and was to be discussed by a planned National Conservation Strategy Conference. This never took place, because the regime collapsed a few months later.

Somalia

As with Ethiopia, and for the same reasons, Somalia's efforts in the field of environmental security, prior to the collapse of the state in 1991, were largely nominal. No central coordinating body charged with environmental protection existed. Several ministries and state agencies were concerned with protection and management of the environment as part of their function. A National Parks Agency was established in 1970 for the purpose of establishing parks and reserve areas, and preparations were made to declare the Las

Badano National Park; but there was not a single protected area listed in the country as late as 1991 (UNEP, 1993, p. 198). The National Range Agency, founded in 1976, was empowered, inter alia, to establish grazing and drought reserves, and to prevent and control soil erosion on the range. An Agricultural Development Project, enacted into law in 1979 for the north-western region, was designed to prevent the desertification of agricultural land in that region. The Ministry of Fisheries, founded in 1977, was responsible for preventing pollution of the sea. The Law on Fauna (Hunting) and Forest Conservation (1969) regulated hunting, and prescribed certain areas to be declared game and forest reserves. A decree issued the same year prohibited the killing of wildlife for the purpose of trophy export.

Among the limited range of concrete steps taken was the prohibition in 1969 of charcoal and firewood export, in order to protect trees. This was amended in 1972 to give a monopoly of charcoal exports to the National Commercial Agency. Other measures were the ending of well drilling, in order to protect pasturage, and the prohibition of sand removal from coastal dune areas. A campaign for dune stabilization launched in the 1970s was a success, although it ultimately stabilized only a small section of the affected area.

Another ambitious project undertaken by the military regime that ruled Somalia from 1969 to 1991 was the resettlement of drought-stricken pastoralists in the wake of the famine in the mid-1970s. Seeking to ease the pressure on the exhausted rangelands, the regime decided in 1975 to resettle more than 100,000 pastoralists who were stranded in refugee camps. Some 15,000 of them were resettled in three fishing settlements – one in northern and two in southern Somalia – and 105,000 in three agricultural settlements along the Shebeli River. The agricultural settlements were managed by the Settlement Development Agency, and the fishing settlements by the Coastal Development Project under the Ministry of Fisheries. Subsequently, these settlements lost nearly half their population, and remained permanently dependent on state subsidies for their survival.

Djibouti

Several areas of environmental deterioration have been identified in this tiny state. They include the degradation of shrubland, woodland and mangroves, salinization of groundwater and soils, sedimentation of coral reefs, small but frequent oil spills in the narrow sea-lane, pollution around the towns of Djibouti and Tadjura, and a reduction in the number of animal species. Hunting for sport, trophies and sale, and habitat degradation threaten many animal species.

Responsibility for environmental protection is shared among several institutions. They include a National Commission for the Preservation of Submarine Fauna and the Environment, the Service of Agriculture and Forests in the Ministry of Agriculture and Rural Development, the Department of Livestock and Fish in the same ministry, the National Tourist

Office, and the Institut Supérieur d'Etudes et de Recherches Scientifiques et Techniques.

Djibouti is party to several international agreements related to protection from marine pollution, and is the site of a marine anti-pollution centre established by the International Maritime Organization in 1963. Two areas, at Musha and Mascali Sud, have been declared marine protected parks.

7.2 Regional Efforts

The Inter-Governmental Authority on Drought and Development (IGADD) *

The Inter-Governmental Authority on Drought and Development (IGADD) was founded on 16 January 1986 by six member-states: Djibouti, Ethiopia, Kenya, Somalia, Sudan and Uganda. It was joined by Eritrea in 1993. The IGADD member-states have common borders and form one bloc, sharing natural resources and socio-economic concerns. Together they cover an area of 5,000,000 square kilometres. Their combined population of about 120 million is the minimum size necessary for sustainable development in terms of optimum use of natural resources, to support manufacturing and industrialization as well as trade, and for the management of a complex communications system. The IGADD countries share a number of important natural resources, such as the Lake Victoria basin, the Nile basin, the Red Sea and the Indian Ocean, as well as many watersheds and mountain ranges. Many communities, particularly pastoralists, live astride national borders that have divided them. On the other hand, the member-states of IGADD had different historical experiences during the colonial period, and have followed varying development strategies after independence.

It was a bold move to combine six countries with varying geographic, economic, social and political backgrounds. Their common denominator was stunted development, which was seen as the main cause of their inability to cope with drought and manage their environmental problems. The move succeeded after the ground had been prepared through massive publicity given to the Horn of Africa by the international media during the height of the famine in the mid-1980s. IGADD has a supreme organ consisting of the heads of states and governments, a Council of Ministers and an Executive Secretariat. The executive secretary, who heads the Secretariat, is appointed by the Council of Ministers for a term of four years, once renewable, and is entrusted with implementation of the resolutions and programmes of action enacted by the Council. IGADD's headquarters are in Djibouti.

*This section combines abridged versions of the contributions made by David S. Maduuli on 'The Inter-Governmental Authority on Drought and Development', and Maina Karaba on 'Infrastructural and Financial Problems of Regional Cooperation in the Horn of Africa', presented at the PRIO/UNEP seminar in Sigtuna, Sweden, September 1991.

The most glaring constraint on regional cooperation in the Horn of Africa is poor transport and communication infrastructure. Some progress has been made over the past 25 years, but much remains to be done. Some 80% of the existing infrastructure was built during the colonial period. The communication axis constructed at that time comprised essentially feeder roads – giant veins that extended from the highly productive areas where raw materials and commodities were produced to the coastal ports for export. The Kenya–Uganda railway was initially the Uganda railway from the Kenyan harbour town of Mombasa to the Kilembe copper mines. The picture is similar for all transport infrastructure in the region: there were no transversal road and railway systems to permit lateral exchanges between countries.

A quarter of a century later, the same infrastructure exists, albeit in a run-down condition. Roads and railways were the first casualties of non-accountability and corruption in the early days of independence. Cases of roads becoming unusable due to disrepair, even before the loans for their construction fell due for repayment, are well documented. Sub-standard infrastructure is not only a shame on African development planners, but also on Africa's external advisers who colluded with politicians and technocrats. The construction of an adequate elementary network of roads and railways would, in itself, create good opportunities for employment, establish new sources to satisfy demands for food and other basic necessities, and attract population to areas of high economic potential. Infrastructural development is investment-intensive, is related to employment and the creation and distribution of resources, and has a direct impact on desertification control and drought management.

Accordingly, in the first IGADD plan for national projects, almost half of the budget was for infrastructural development while, among the sub-regional projects, three-quarters were intended to strengthen communication. Unfortunately, plans are one thing, reality is another. None of the aid sought for infrastructure was granted, and a proposed consortium of donors to fund infrastructural development never materialized.

IGADD developed a food strategy for the region, which was endorsed by all member-states in Kampala in October 1990. Prior to this, a food security study was carried out for IGADD, and a report was issued in 1990. IGADD's food strategy established four priority areas: increasing food production, improving food marketing, increasing food consumption, improving food policy management. As a first step, a regional early-warning food system project was established in Djibouti to serve all six countries. A sensing component of that system is in Nairobi. The goal of this project is to develop a databank covering the entire food security spectrum. Till then, IGADD must take its information from member-governments, without being able to verify it.

Regional cooperation to prevent soil erosion*

In the Horn, both the arable highlands and the pastoral lowlands face mount-
ing environmental stress. As a consequence of demographic transition, the
traditional crop and livestock production systems no longer appear sustainable. Part of the problem is population growth. The overwhelming majority of
the population consists of primary producers growing crops and herding livestock. Their production technology was developed to suit a situation where
land was plentiful, people were few and mobility was easy. When yields
declined, the people moved, and soil fertility was restored during long fallow
periods. However, the present generation of cultivators and livestock herders
faces a situation of increasing land scarcity. Cultivation has to be intensive,
which means exhausting the land. Value is created by destroying the capital.

Environmental stress is manifested in deforestation, soil erosion and a general desiccation of the landscape – phenomena that are also linked to water
scarcity (Falkenmark et al., 1989). There are various types of water scarcity.
Hydroclimatic water scarcity is due to limited precipitation during a short
rainy season and to considerable inter-annual variations. This is most prevalent in the lowlands, and there is little that can be done about it.
Consequently, the semi-arid and arid lowlands have a limited potential for
supporting agriculture.

The highlands have much more potential, but this may be undermined by
unsustainable farming methods. Falkenmark et al. have pointed out that
chronic water stress is also caused by over-exploitation of the eco-system,
leading to degradation. The water-holding capacity of the soil is thus reduced,
and an increasing proportion of rainwater disappears as run-off, while average soil moisture is reduced. In the northern highlands of Ethiopia the land
is now on the brink of ecological collapse. As a result of intensive cultivation
and grazing, triggered by population increase and an almost total lack of off-
farm employment, the hills are bare and the soil is washed away during heavy
rainstorms.

Environmental rehabilitation programmes need to address the problem of
water availability in the root crop zone, as well as the problem of soil fertility
in general. These problems must be considered at three levels: farm, catchment region and government. At the farm level, remedial activities include
construction of terraces and soil bunds, as well as planting grass strips along
the contour, mulching, digging infiltration ditches, etc. Roaming livestock
must be controlled and gradually modified to zero grazing and fodder production systems. At the catchment level, large-scale activities must be
undertaken: reforestation of mountain slopes, gully plugging, construction of
rock catchments to collect excessive run-off water, closing vulnerable slopes to
grazing and cultivation. Such technical measures alone are not sufficient. At

*The following two sections are taken from Michael Stahl's contribution on 'Regional
Cooperation to Prevent Soil Erosion in the Horn of Africa', presented at the PRIO/UNEP seminar in Sigtuna, Sweden, September 1991.

the government level, an enabling environment for long-term investment in land improvement must be framed through local participation, secure land and tree tenure arrangements, a pricing structure that motivates farmers to invest, technological improvements through research, and effective extension services.

The land tenure system is a critical determinant of people's willingness to invest labour and money in land improvement. Throughout Africa, rules and legislation related to land ownership tend to be diffuse and fluid. In areas where customary tenure prevails, people are seldom motivated to engage in soil conservation, preferring instead to look for new land when their plots are exhausted. In Ethiopia, the socialist experiment thoroughly alienated the peasantry. By contrast, individual titles to land in Kenya have been instrumental in creating a favourable attitude to conservation. Soil conservation on farms has taken place mainly in areas where private ownership prevails. This is not to suggest that complete commoditization of agricultural land would create maximum incentives for soil conservation. Nevertheless, land-tenure aspects of sustainable agriculture are extremely important and merit further study.

Scope for regional cooperation

In the two countries with the longest experience of conservation work, Kenya and Ethiopia, specialized divisions for conservation have evolved within the ministries of agriculture. Since the institutional setting and the policies differ between countries, it would seem there is limited scope for regional cooperation in conservation. However, regional cooperation could be considered in border areas with similar ecological and socio-economic characteristics. The most obvious case would be the rangelands in the semi-arid belt bordering Uganda, Sudan, Kenya, Somalia and Ethiopia. There is potential here for agreements allowing border crossing by livestock and herders, joint utilization of water points, etc. An activity of immediate significance would be to stop the rampant cattle rustling across borders which, in recent years, has become extraordinarily violent, leaving many areas insecure. Locust control is another field that would benefit from regional cooperation.

There is also considerable scope for regional cooperation in research and technology development, postgraduate research training, as well as specialized training for extension and specialist staff. Knowledge of the region's various agro-ecological zones is still limited. Much research remains to be done in order to develop resilient production systems, given the trend towards intensification of land use. Research on soil conservation has started in all countries, but it is hampered by staff and resource limitations. The International Council for Research in Agroforestry (ICRAF), with headquarters in Nairobi, conducts and coordinates research on the integration of woody perennials into farming systems within different agro-ecological zones. A network has been set up for the highlands of East and Central Africa, which have a bimodal rainfall pattern. Kenya, Uganda, Rwanda and Burundi are members of the network, and Ethiopia has been invited to join.

Very few researchers are involved in environmental research at postgraduate level in the region. Therefore, regional cooperation is a prerequisite for the creation of a critical mass of researchers and the generation of research findings of a scale sufficient to have an impact. By exchanging available expertise and cooperating in developing a corps of professionals, the countries of the region would have a lot to gain. First of all, it is cheaper to engage experts from the region than to hire consultants from the industrialized countries. Second, since the environmental situation in most countries of the region has many similarities, regional experience can be utilized at home. Most important, regional exchange of experience can become the cornerstone for long-lasting institutional cooperation. It is now acknowledged that Africa's development problems cannot be solved by massive capital injections alone. A crucial component is the emergence of local scientific and technological capacity that can adapt international research findings, technologies and policies, as well as productive indigenous knowledge relevant to the problems facing the region.

While there may be various problems – political and other – in promoting regional cooperation, the most common one is lack of necessary foreign exchange. The immediate function of donors is to provide this kind of support, with the hope that regional networks will develop and become sustainable in the future. For this to succeed, it will require political commitment on the part of the states involved, and cannot rely forever on foreign donors alone. However, given the political instability and economic decline faced by the countries in the region, it would be overly optimistic to expect that soil conservation and environmental rehabilitation projects will have priority in state allocation of scarce budget resources. The sobering conclusion is that large-scale programmes, including projects for regional cooperation, will continue to rely on external donor funds throughout the 1990s.

Locust control*

Locusts have menaced human livelihood in tropical and sub-tropical areas since time immemorial. As migratory pests they know no frontiers, but move in swarms over large areas, devastating crops and vegetation in their wake. In the past 50 years, the response to this threat has been intensive research into locust biology, swarm dynamics and the development of new control methods. Several national and international research and control organizations have been created, and the FAO has become the main coordination agency for control measures and resource distribution.

The desert locust (*Schistocerca gregaria*) is the predominant species in the Horn of Africa. What distinguishes locusts from other species of *Acrididae* (grasshoppers) is their occurrence in different phases, ranging from the solitary

---

*This section is an abridged version of Stephan Gunnarsson's contribution on 'Inter-state Cooperation in Locust Control in the Horn of Africa', presented at the PRIO/UNEP seminar in Sigtuna, Sweden, September 1991.

to the gregarious. In the solitary phase, locusts show no tendency to aggregate and are found in low densities in certain recession areas. As the normal lifespan is six months, two or three generations can be produced each year, and survival depends on climatic conditions; rain means access to food. If there is abundant rain for several consecutive seasons, breeding and growth conditions become favourable and more generations are produced. Increased densities lead towards a gradual transformation to the gregarious phase. Behaviour changes, with a tendency to aggregate and to migrate in swarms that take off in the morning and fly all day until dusk. Due to the long time they spend flying, migrating locusts need a substantial intake of energy and consume approximately their own weight each day, eating virtually anything green.

The range of the desert locust is from the west coast of Africa to India in the east, and from the Equator in the south to 40° North: an area of nearly 30 million square kilometres. However, it is only during plagues that swarms invade all of this area. The solitary locust population is restricted to the so-called recession areas, found throughout the Horn of Africa. In climatically normal years, without heavy and prolonged rains, increases in locust densities sometimes occur in the recession areas, leading to local outbreaks where gregarization is initiated. The small swarms that result from this will move with the wind towards new rainfall areas where further breeding can take place. Depending on weather conditions, these local outbreaks may dissipate, or turn into plagues.

Eastern Africa has been invaded by plague-scale swarms several times this century, on average once every decade. The 1940s and 1950s were a period of more or less constant plagues. The beginning of the 1960s saw a sharp decline in swarms, and populations remained low up to 1967, when a new upsurge commenced. This led to a short-lived plague in 1969. During the 1970s, there were scattered upsurges, but no plagues developed until 1977–79, when densities increased to plague dimensions. During the late 1980s, there was a very large plague all over northern Africa and the Middle East.

Until the advent of organic chemical insecticides in the late 1940s, there were no really effective means of dealing with locust swarms. In addition to futile attempts by peasants to scare off the locusts by shouting and beating the vegetation, bran mixed with arsenic was often used; a bait which the hoppers ate readily. However, besides the danger to the handlers of this poison, there were great difficulties in transporting and spreading the bait. New chemical insecticides were more effective in killing the locusts, but they could be used on a large scale only when appropriate methods for spraying from the air were devised. New methods in estimating the size of locust populations before and after control have also been of great importance. Controlling swarms when they appear is not enough, since a small swarm of only one square kilometre can consume around 100 tons of vegetation a day. A forecasting system is extremely important. Such a system relies mainly on weather information that can be used to locate local outbreaks and predict locust movement. Radar can be used to trace flying swarms and to guide aircraft in locating and following locust populations.

Up to the 1980s, all modern control methods relied on chemical insecticides. Several types were used, with varying degrees of toxicity and persistence in nature. The most effective ones were also the most persistent and toxic, and thus liable to have adverse effects on humans and animals. The most effective one, Dieldrin, became the most controversial. It was banned in several African countries, and some donor countries refused to finance its use. As a result, research has now turned to biological methods of control using bacteria, fungi and viruses (Panos Institute, 1993).

It was the colonial powers that initiated most of the control and survey work in Africa, such as in Eastern Africa. After independence, regional organizations were founded, such as the Organisation Commune de Lutte Anti-Acridienne et de Lutte Anti-Aviaire (OCLALAV) in West Africa, and the Desert Locust Control Organization for Eastern Africa (DLCO-EA). The latter was founded in 1962 with Ethiopia, Djibouti, Somalia, Kenya, Tanzania, Uganda and later Sudan as members. In addition to funds provided by the member governments, a substantial part of the DLCO's expenses has been covered by bilateral and multilateral aid agencies, mostly channelled through the FAO. The DLCO has been responsible for organizing and carrying out surveillance and control throughout the region, and to some extent also on the west coast of the Arabian Peninsula.

Unfortunately, both organizations lost much of their operational capacity during the 1980s. This was due mainly to the unstable political situation, but also to a decrease in incentive and alertness which followed upon the successful control efforts of the preceding years. Member-states failed to pay their annual contributions, and the organizations were unable to maintain their personnel, equipment and stores of insecticides. The DLCO-EA could not even pay its staff for several months during 1993, and a fund-raising drive was launched to resuscitate the organization. During the most recent plague, most control operations were organized directly by FAO, which to a large extent used commercial aircraft, since the fleet of the control organizations was quite inadequate.

Rinderpest*

Rinderpest, the greatest cattle scourge ever known, has had a devastating effect in Africa. Although vaccination effectively brought the infection under control, particularly after a concerted campaign launched in 1962, foci of infection were left untouched. That, together with the relaxing of continuous surveillance and control actions, brought about a resurgence of this plague in the late 1970s and early 1980s in the Sahel and the Horn of Africa. The OAU (Organization of African Unity) and the Inter-African Bureau of Animal Resources launched a new campaign, the Pan-African Rinderpest Campaign,

*This section is an abridged version of the contribution on 'Rinderpest Control' presented by Omar Sheikh Abdurahman and Set Bornstein at the PRIO/UNEP seminar in Sigtuna, Sweden, September 1991.

designed not just to control but to eradicate the disease from Africa permanently. The rinderpest story illustrates very well the vital importance of regional and international cooperation in fighting animal disease.

Rinderpest is a fatal virus infection of ruminants and pigs. It probably travelled with cattle-drawn supply caravans of traders and armies invading Europe from Asia. It came to Africa in 1842 with cattle imported into Egypt from Romania, and spread as far south as Khartoum with the British campaigns in Sudan (1884–85). When the Italians invaded Eritrea in 1889, rinderpest was carried with cattle brought from Aden and India. The infection spread rapidly in a population of animals completely lacking immunity. It swept across East Africa, crossed the continent from east to west, and nearly reached the Cape in South Africa. It killed perhaps 90% of the cattle in its path. Famine followed, decimating the population. The scourge altered the political and social balance, breaking the economic backbone of prosperous and advanced communities, and initiated the breakdown of a long-established ecological balance. Large tracts of land were depopulated of livestock and consequently also of people, leaving the soil untilled and pastures ungrazed. Such land was opened to the invasion of bush, which was then populated by wildlife. The return of the game and eventually also cattle provided the biomass suitable for tsetse flies, vectors of sleeping sickness and trypanosomiasis.

Before the advent of modern vaccination, herders in Somalia used to avoid areas where rinderpest occurred. They also attempted to immunize their cattle by diluting the urine, faeces and milk from infected animals and instilling a few drops in the nostrils of healthy ones. During the pre-vaccine era, it proved possible to eliminate the disease from Europe, Japan, Southern Africa, Australia and Brazil, by restricting the movement of cattle and destroying sick animals as well as those that came into contact with them.

The development of a vaccine in 1931 raised hopes of eradicating the disease. Vaccination campaigns began in Kenya and Somalia as early as the 1940s. Since 1957, a safe vaccine which gives lifelong protection has been used in many countries, with good results. However, there is always the risk of re-infection across national borders. This frustrating situation brought about international cooperation in fighting the disease. In Africa, a mass vaccination campaign called the Joint Project 15 (JP15) was launched in 1962. The campaign was concluded in 1976.

Though effective in reducing the incidence of the disease in the short term, JP15 was a failure in the long run, because not all countries maintained follow-up measures. Some countries preferred to believe rinderpest had been eradicated permanently, while others were forced by economic necessity to cut budgets. Political conflict elsewhere prevented investigation and surveillance of local outbreaks. Rinderpest has long been associated with unrest, migration and wars, events that upset quarantine and surveillance measures. Resurgence of the disease was reported in the late 1970s beginning in West Africa. Mauritania and Mali reported outbreaks in 1978 and 1979 respectively. The infection spread eastwards along cattle trade routes, reaching the

Horn in the early 1980s. Two to three million animals succumbed during the first wave of this pandemic.

This resurgence of rinderpest made governments aware of the necessity to rid the world of this scourge once and for all. In 1980, at a meeting in Rabat, it was decided to launch the Pan-African Rinderpest Campaign. Six years passed before final decisions were taken and agreements were signed in Addis Ababa in 1986. One year later, the first technical committee meeting was held in Nairobi. This seven-year delay cost at least 3 million head of cattle. The largest multilateral donor was the EEC, and many bilateral donors were involved in the implementation of the programme in individual African countries. The FAO supported the campaign by providing vaccine quality-control laboratories, equipment and training for vaccine production. Vaccine banks were set up in five countries, where vaccines were available at short notice. Part of the funds for the programme were earmarked for research on developing more heat-stable vaccines, on the role of wildlife in the transmission of rinderpest, and on the immunosuppressive action of the rinderpest virus. Mindful of the JP15 experience, the intention of the new programme was to strengthen national veterinary services to ensure follow-up measures.

Of the 30 countries participating in the Pan-African Rinderpest Campaign, only Sudan, Ethiopia and Somalia have experienced outbreaks of the disease, due to political conflicts which prevented implementation of the campaign. In Ethiopia, one spray airplane was shot down and another was fired on in 1987. The embattled areas of Eritrea and Tigray were not accessible from Ethiopia, and the Ethiopian government refused to give permission for operations in these provinces to be carried out from Sudan.

Animal disease control needs to be harmonized among neighbouring countries. Nomadic grazing grounds are vast, whereas veterinary services are usually restricted to the vicinity of towns and villages. It is important to establish health services along dry-season grazing areas and in frontier areas on both sides of national borders. Bilateral projects for the control of animal disease should focus on establishing health facilities on transboundary routes. The implementation of the Pan-African Rinderpest Campaign is breaking new ground in inspiring cooperation between governments and international agencies, as well as between communities and individuals within countries. The OAU member-states are on their way to achieve control and perhaps eradication of rinderpest by the year 2000.

7.3 International Efforts

The Nile River basin

Taken as a whole, the Horn is a region of water scarcity. This situation is likely to worsen in relation to the needs of a fast-growing population, and even countries with relatively high precipitation such as Kenya and Ethiopia are likely to be at risk in the not too distant future (Falkenmark, 1987). Two factors may aggravate the situation. One is the climatic trend that has reduced precipitation throughout the region; the other is the additional demand for water envisaged in the development plans of most countries in the region. Ethiopia and Kenya can attempt to solve the problem through engineering schemes designed to conserve water that falls on their highlands by damming rivers, taking water out of lakes and generally restricting run-off. If at all successful, such schemes will vitally affect downstream countries like Sudan, Egypt and Somalia, and may provoke conflict.

There is scant precedent for regional cooperation in this vital area. The 1959 accord between Egypt and Sudan for the 'full utilization of the Nile Waters' was reached after several years of acrimonious negotiations. Egypt's preference for a giant dam at Aswan devoted to its own needs was initially rejected by Sudan, whose government preferred a radically different scheme based on an integrated approach in which other riparian states would utilize a series of smaller, inter-dependent conservation and storage projects (Khalid, 1984). There was also disagreement over the share of water proposed for Sudan. Talks broke down, and it was only in 1959, following the first military takeover in Sudan in 1958, that agreement was reached on Egypt's terms. At that time, Egypt also concluded an agreement with Uganda to build a dam at Owen Falls in Lake Victoria, with the dual purpose of controlling the flow of water from the lake and producing electricity for Uganda.

Subsequently, both Egypt and Sudan sought to promote regional cooperation. The terms of reference of the Permanent Joint Technical Committee, formed by these two countries as a result of the 1959 agreement, empowered the Committee to liaise with other riparian countries – including Ethiopia, Kenya, Tanzania, Uganda, Rwanda, Burundi and Congo (Zaire). In 1960, the Committee invited Tanzania, Uganda and Kenya to send representatives to its meetings. In 1963, following a rise of 2.5 metres in the level of Lake Victoria and widespread destructive flooding, Uganda, Tanzania and Kenya joined Egypt in requesting international support for a survey of the catchment area that feeds Lakes Victoria, Keyoga and Mobutu (Albert). The so-called Hydromet project eventually got under way in 1967. The preamble to the agreement for the survey, signed by all Nile riparian states except Ethiopia and Congo (Zaire), stipulated that the project 'provides the groundwork for inter-governmental cooperation in the storage, regulation and use of the Nile'. The Hydromet Technical Committee was set up in 1973 by the same group of countries, with Ethiopia and Congo (Zaire) as observers, and UNDP backing, to collect hydrological data in the Upper Nile basin.

In 1978, having concluded the deal to construct the Jonglei Canal in Southern Sudan, and concerned about water availability in the future, Egypt and Sudan proposed setting up a Nile Basin Commission to be entrusted with the tasks of conducting hydro-meteorological studies, establishing data banks, sponsoring studies pertinent to the conservation of the Nile waters, sponsoring studies of river control, preparation of working arrangements of dams and standardization of hydrological equipment and methods of measurement. This did not bear fruit. In 1983 Egypt hosted a meeting in Cairo attended by representatives of all the riparian states, except Ethiopia. The group, which was called *Endego* from the Swahili word meaning 'brothers', held several meetings at ministerial level to discuss issues of regional interest, including the management of rivers and lakes.

In 1982, the UNDP proposed to expand the scope of Hydromet activity. In 1986, it invited the Nile riparian states to attend a meeting in Bangkok hosted by the Secretariat of the Committee for Co-ordination of Investigation of the Lower Mekong River Basin, to assess the experience of that organization. There was unanimous agreement that joint management of the Nile River is essential, and a request was submitted to UNDP to carry out the required studies. The Ethiopian ambassador in Paris attended the Bangkok meeting and associated himself with the consensus, but without committing his government. In 1987, a UNDP pre-identification mission visited all the riparian states, except Ethiopia, to seek approval of the steps to be taken towards the identification of concrete projects and to plan for a meeting at a ministerial level. In 1989, a UNDP fact-finding mission visited the same countries to identify their specific interests in regional projects and to assess their needs. It submitted a report on the water and energy needs of these states, and recommended the formulation of a programme for regional development.

Once more, Ethiopia did not participate. Its government objected strongly to the terms of reference of the fact-finding mission that had been drawn by the Hydromet Technical Committee. Ethiopia's objections were many, and most of them related to the assumption that the water needs of the downstream countries – Sudan and Egypt – are greater than the needs of any other riparian state, that these two countries have priority in the use of the Nile water, and other states must take these into account when they formulate their own plans. Ethiopia, which by then had established an Ethiopian Valleys Authority and was drawing up ambitious irrigation plans of its own, once more proclaimed its sovereign right to utilize the natural resources within its borders, and stated its willingness to cooperate in joint projects only on the basis of 'equitable and rightful entitlement' to the use of Nile waters. The UNDP initiative was then abandoned.

*External aid**

Continued political instability and economic stagnation in Africa have modified the philosophy of aid offered by the industrialized world. New criteria have been introduced as essential conditions for sustainable development. Among these are good governance and sound management of the environment. The revised Norwegian aid policy is an example of efforts made to fulfil these new criteria.

The Norwegian Agency for Development Cooperation (NORAD) has discussed ways of redefining its donor role. The starting-point has been the

*The data for this section are taken from Espen Wæhle's contribution on 'The Norwegian Donor Perspective on Regional Environmental Cooperation in the Horn of Africa', presented at the PRIO/UNEP seminar in Sigtuna, Sweden, September 1991.

realization that many developing countries have become dependent on a form of development assistance whose terms are strongly influenced by donor preferences. As a result, aid has not been integrated into the political, economic and administrative structures of the developing countries. This situation must change. To that end, the current NORAD (1990) strategy contains three key concepts: responsibility, accountability and good governance. NORAD's main task is now to provide assistance to developing countries in their efforts to bring about lasting improvement for their entire population. This concept of 'assistance' implies that Norway is able to contribute on a limited scale, while main responsibility for development in a particular country lies with that country itself. The concept of 'lasting improvement' indicates a desire to contribute to development that can be sustained without external assistance. Finally, the development process has to be based on the ecological, cultural and socio-economic conditions prevailing in the recipient countries.

NORAD does not intend to 'plan' projects and programmes in the future. Instead, NORAD will increasingly emphasize the development and strengthening of institutions, the clear demarcation of roles for NORAD and cooperating partners, and the adoption of long-term objectives. It will require recipients of Norwegian aid to follow up projects for which they are responsible. NORAD cannot delegate responsibility without requiring accountability; this in turn means good governance. Good governance includes the democratic means of assuring accountability to both the people of the country and its partners in development. This brings up the delicate question of how to apply these basic principles to development cooperation, principles which include 'a concern for human rights and democracy, ecological sustainability, the environment, women's rights and the goal of an egalitarian society . . . These are not moral dictates. In our perception they are prerequisites for development that will improve the lot of the people' (Grimstad, 1991, p. 4).

The Sudan–Sahel–Ethiopia (SSE) Programme may serve to illustrate some of the issues related to Norway and its environmental cooperation in the Horn of Africa. During the drought and ensuing emergency situation in 1984–86 in the Sahel countries, the Norwegian government supported various relief operations across the African continent. As the emergency came to an end, the Norwegian aid administration started to develop plans for long-term assistance to the region. It was widely recognized that famine was part of a deep structural crisis and that good rainfall was not the sole factor that could radically improve the situation. External and internal causes were involved; political, economic and natural factors had all contributed to a range of problems.

The Norwegian government committed itself to continue providing relief assistance, but it was evident that this was insufficient. In order to move beyond mere relief operations in the Sahel, Norway decided to expand its long-term development cooperation through a special Sahel programme designed to address the problem of food security and to improve the ecological basis, in order to develop sustainable production systems. Sudan,

Ethiopia, Somalia and Djibouti are among the countries included in this programme. Three disbursement channels were singled out for this programme: international and/or multilateral institutions, Norwegian non-governmental institutions and their partners in the South, and research institutions, including Norwegian ones and their counterparts in developing countries. The Programme was initially intended to cover a period of five years, with a proposed budget of one billion Norwegian kroner (USD 130 mill.). To increase the effectiveness of the programme, financial support channelled through NGOs and the research component were concentrated in Sudan, Ethiopia and Mali.

The Nordic countries meet each year to discuss matters of common interest in the Sahel, including the Horn of Africa. Denmark, Finland, Norway and Sweden have all supported activities in the Horn, though they have worked through different channels and partners, and in different sectors and types of projects. Denmark, Finland and Sweden have used multilateral organizations, especially the UN Sudano-Sahelian Office, while Norway has worked with multilaterals, NGOs and research institutions. Compared to other regions where the Nordic countries have assisted in development, there is little project and programme overlap in Sahelian activities. In recent years, however, all the Nordic countries have withdrawn support from some countries in the Horn, a reflection of Nordic concern for human rights and, in a broader sense, for good governance.

7.4 Conclusion

Some portions of the Horn's natural endowment are in danger of depletion; meaning that the ability to perform their natural function is threatened. Struggling to meet current needs and distracted by violent conflict, societies and governments in the region have not been able to deal meaningfully with this threat. Nor is it likely that, on their own, they will be able to do so in the foreseeable future. In fact, the countries of the Horn, like many others in Africa in the last decade of the 20th century, no longer manage their resources on their own. Their freedom of action has been considerably limited by the structural adjustment programmes (SAP) designed by the World Bank and the International Monetary Fund to improve economic performance in the continent. Adherence to the rules of SAP has become a strict condition of eligibility for further borrowing and aid, as Sudan discovered when it was disqualified for failing to do so. Western oversight has not been limited to the economic sphere. Conditionality has extended to the political sphere, where it has aimed to encourage democratic reform in the post-colonial state.

Given the limited capacity of the states in the Horn to address environmental and other problems on their own, the regional perspective has gained favour, particularly among international agencies and foreign states providing aid to the region. The Inter-Governmental Authority on Drought and

Development (IGADD) was founded in 1986 amidst great expectations. It was intended to develop a regional approach and to design projects on that level, but failed to do so, compiling lists of national projects instead. It disappointed its backers, and its first two executive secretaries were changed after a single term in office. IGADD served a different purpose when it became the venue for meetings of heads of states in the region, who found it a convenient forum for negotiations that had little to do with IGADD's actual mandate. In the mid-1990s, IGADD served as a forum in the search for peace in Sudan. An attempt was made in 1996 to re-launch the organization, now renamed the Inter-Governmental Authority on Development (IGAD), with a revamped structure and a new executive secretary.

The threat to environmental security in the Horn of Africa will not receive the attention it merits as long as absence of peace and political stability remain the main life-threatening conditions in the region. It is not surprising that Sudan's (1990) *National Economic Salvation Programme, 1990–1993* contains no reference to the environment. It is to be expected that conditionality by donor countries and agencies will be extended to require some protection of the environment in the receiving countries. However, it seems doubtful whether this will produce anything more than merely formal acquiescence on the part of the latter.

8
What Can Be Done

8.1 Introduction

In the pages that follow some of the main issues examined in previous chapters are put into an analytical context appropriate to the region and the historical phase it is going through. From this comes the realization that approaches to the region's many problems cannot be piecemeal or simply technical, but must involve fundamental changes in the structures and policies of the post-colonial state.

8.2 Pastoralism

The historical reasons for the decline of pastoralism were described in the first part of this work, while the epidemic of conflict that has swept the region and accelerated the process of decline was treated in the second part. A basic factor in the marginalization of pastoralism has been the complete disregard of it by ruling groups and development experts. Such disregard was premised on the assumption of economic irrationality attached to traditional livestock production by earlier anthropologists, and popularized with the slogan 'cattle complex'. Pastoralism was seen as a primitive mode of subsistence, irrational in economic behaviour because it is not market-oriented, and incapable of development within its own framework. Furthermore, pastoralists were regarded as the scourge of the environment. 'The nomad is not so much the son of the desert as its father', asserted scathingly a Sudanese official report (Abdel Ghaffar, 1976, p. 137; see also 1987), while a Somali government publication (Somalia, 1977, p. 29) declared 'the part of society which impedes economic prosperity is the nomadic population which raises only livestock'. This was at a time when livestock exports from the nomad sector were Somalia's main foreign exchange earner.

By contrast, a later generation of anthropologists and specialists has come to regard traditional pastoralist production techniques as highly sophisticated, honed through age-old experience to be the most suitable for using and preserving the scanty resources of the arid environment. It is now understood that pastoralist reluctance to sell animals is due not to an irrational 'complex', but to highly rational calculations relating to a host of objective factors: the maintenance of adequate food safety margins in unpredictable circumstances, the forging of strong social ties and attaining social status within the community, the inaccessibility of market outlets, the usually negative terms of trade, and the lack of investment possibilities for surplus finance capital. It is also realized that traditional practices, such as uncontrolled stock proliferation, have become untenable in the congested circumstances of the present and must be modified.

Such knowledge has made no impact on policy-makers and their advisers in the Horn. They remain determined to eliminate traditional pastoralism by encouraging or compelling nomads to 'settle down'. Sedenterization has proceeded apace in the region, promoted by impoverishment, political insecurity and increased dependence on food provided by relief agencies. This type of 'settlement' has occurred in peri-urban slums, refugee camps and around food-relief stations, where destitute pastoralists have swelled the ranks of the unproductive and dependent population. The negative attitude of ruling groups is based on two additional considerations. One concerns the potential uses of land in the pastoralist habitat. There is hope that part of this land can be turned to cultivation, through irrigation or otherwise, and thus could help ease population pressure in other areas and increase food supply. However, there is no evidence yet that land in the arid region actually can produce more, or even as much, under a different mode of production, and the results of experiments carried out thus far have not been encouraging.

Alternatively, these areas could be given over to meat production through modern ranching schemes. A number of such schemes, based on the North American and Australian models, have been established in the Horn. Designed to take advantage of free land in the pastoralist habitat, cheap labour from the pastoralists themselves, and young animals from pastoralists' herds to be fattened and sold, they were intended to drain the remaining resources from the pastoralist domain in order to provide cheap meat for the domestic and export markets. The performance of these schemes so far has not been encouraging. Planners and policy-makers in the Horn have been interested in the land over which pastoralists roam – not in the pastoralists themselves. This interest was manifested also in the so-called 'rangeland development' programmes adopted by all the countries in the Horn. These were primarily designed to increase the livestock carrying capacity of land through range management, water development, road building, etc. Alternative uses of the pastoralist habitat like tourism, developing geothermal energy sources, etc., are often in the forefront of official interest. Since the intention is to develop the land, not the pastoralist mode of production, the objectives of planners do not always coincide with the interests of the pastoralists.

The second consideration of ruling groups is political. In most instances, pastoralists inhabit the periphery of the state, maintaining close contacts with kin groups across the borders and showing little respect for border restrictions. Moreover, they have a long history of smuggling high-priced goods, including weapons which they often use to defy state authority. Consequently, officialdom regards the wandering herders as security risks and potential subversives. The fact that the latter frequently do fulfil such expectations often makes them victims of state violence.

The decline of pastoralism and the conflict it has endangered constitute a human problem of the first magnitude. This has been recognized by the United Nations, which in 1990 established a short-lived project for Nomadic Pastoralists in Africa (NOPA) headquartered in Nairobi, to identify problems and study possible solutions. Policy-makers must be encouraged to accept the fact that, with the given resources and the current state of technology, the arid lowlands can be effectively exploited by their present inhabitants, provided they are assisted to reverse the long trend of decline. This must be done in the context of an agrarian reform focused on the pastoralist sector. Various types of agrarian reform have been carried out in the states of the Horn, but none of them involved the pastoralist sector even marginally. Indeed, often the effect of land reform was to undermine further the land tenure rights of pastoralists. It is a glaring fact that pastoralists have no secure rights over land in any of the countries in the Horn. All land in the pastoralist realm is state domain, and can be alienated without even informing its inhabitants.

A programme of environmental protection and development must be designed to reverse the process of degradation in the arid zone. This must concur with the wishes, and serve both the short and the long-term interests, of the people who live in this zone. The time-span in which people perceive their own interests is determined by their circumstances; if they seem heedless of the distant future, often it is because they are preoccupied with present survival. The failure of most development projects designed for pastoralists in the past is another glaring fact which bears a message for the future. A development programme for the pastoralist milieu must provide for mobility – an imperative of pastoralist production that has been steadily curtailed during this century. Mobility is essential not only for production, but also in order to integrate the pastoralist economy into the regional market for the benefit of wider communities. Experience shows that when the market is accessible and the price is right, pastoralists are not averse to disposing of surplus stock. This was the case in the Horn during World War II, when food and pack animals were bought in pastoralist areas by the British and Italian armies.

State frontiers are often an obstacle for pastoralists on the way to their natural markets. The markets for the Ogaden in Ethiopia are in northern Somalia, for the Ethiopian Afar in Djibouti, for the Eritrean Beni Amer in eastern Sudan, for Southern Sudanese in Uganda and Kenya, etc. State borders violate the logic of the natural economy, and impede precisely what development planners seek to promote: pastoralist integration in the market. State and provincial borders also impede the process of natural adjustment that balances people, animals and land and is a key feature of ecological adaptation and survival of traditional pastoralism. Finally, state borders impede contact between kinsmen, and between people and their religious and ritual leaders who may live in another state. Understandably, pastoralists are forced to ignore state borders, but not without economic and political consequences. Trading in this manner becomes smuggling and incurs additional costs and hazards, while disregard for 'national' borders reinforces official wariness of pastoralists and results in frequent violent confrontations.

Restoring a measure of pastoralist mobility will require imaginative appropriate political arrangements between and within states in the Horn of Africa. Easing border restrictions and removing imposts on local commerce would greatly encourage trade. In the early 1990s, when such restrictions lapsed due to the weakening of state control, unofficial cross-border trade flourished, while official interstate trade was non-existent. Relaxing restrictions would enable pastoralists to make better use of regional resources, and ease pressures on congested areas, with a beneficial environmental impact. It would also help to ease the burden of separation of kin groups imposed by colonialism, and thus redress a historic injustice.

Resolving the plight of pastoralists in the Horn will require interstate co-operation and firm political commitment. A regional agreement, a sort of 'Pastoralist Covenant', is needed to define a special status for the pastoralist realm, to plan for its recovery and development, and to protect the herders on the move, especially across state borders. A regional research institute is needed to provide the knowledge required for planning the rehabilitation of the pastoralist habitat, and to look for solutions to such intractable problems as providing health, education and other social services for herders on the move. A draft agenda for such action was prepared by the Project for Nomadic Pastoralists in Africa in 1992 (NOPA, 1992).

Pastoralist interests need to be articulated and defended by the pastoralists themselves through their own political organizations. Quite obviously, as Sandford (1976, p. 4) has noted, 'unless pastoral societies are politically important and well organized to express their own points of view, pastoralist objectives will be neither well understood nor much heeded in the planning of development programmes'. Pastoralists in the Horn have not been politically inactive. On the contrary, as shown in Part Two, their alienation is reflected in heavy involvement with movements defying the authority of the post-colonial state. By the early 1990s, state authority had collapsed in Somalia and Ethiopia, and was on the verge of collapse in Sudan and Djibouti: the pastoralist contribution to this state of affairs was not insignificant. Clearly, state structures in the Horn need to be transformed, and adequate recognition and representation of pastoralist societies ought to be one of the features of the future state. If not, there is bound to be conflict without end in the region.

8.3 African Traditional Forms of Cooperation*

Is there an established tradition of environmental cooperation in eastern Africa? If so, what forms does it take? Can they be elaborated and used in the context of modern environmental cooperation across the borders of nations and regions? The notion of cooperation implies a conscious, concerted activity

---

*This section is taken from Gudrun Dahl's contribution on 'African Forms and Traditions of Cooperation: Their Relevance and Applicability Today for Regional Environmental Cooperation', presented at the PRIO/UNEP seminar in Sigtuna, Sweden, September 1991.

towards a particular goal: in this case, one which has to do with the environment. In the modern context, it is rarely a problem to define which issues pertain to environmental security. In the traditional and non-Western context, however, this is more difficult. Concepts such as 'nature' and 'environment' have varied meanings, and activities which to us may seem related to the environment are not necessarily thought of in such terms by traditional actors. Should we include only forms of cooperation explicitly involving protective attitudes and actions towards nature? Or can we include also any form of social cooperation relating to our definition of environmental security, whether or not other people think of them that way? Both these perspectives will be considered here, in relation to livestock-rearing peoples in East and Northeast Africa.

Pastoral people naturally maintain ideas about their natural environment, and act accordingly. This does not necessarily mean they have profound systemic insights corresponding to some scientific model of ecological processes. On the whole, pastoralists, like other subsistence producers, tend to have detailed notions about the way many aspects of their biological environment work. In their experience, many of these notions have proved true and efficient, whereas others may be misconceived and counter-productive, while others still may be manifestly wrong yet still serve a purpose.

Do we find goal-directed environmental behaviour in the traditional pastoral cultures of eastern Africa, and can it serve as a basis for local or regional cooperation? The proper care of livestock is the central cultural concern for pastoralists. Certain groups have quite extensive networks for the distribution and management of animals. Is this 'environmental cooperation'? The question of herds illustrates the difficulty of defining the boundaries of 'nature'. Whether or not we include herds in our definition of 'environment' depends on our cultural constructs, as do our ideas of what pertains to 'nature' in general. Obviously, there is an arbitrary boundary between 'maintaining natural resources' and 'domestication'.

Grazing grounds and water resources fit our notions of 'nature' better. The need to conserve and protect such resources is rarely a theme that is culturally elaborated to the same extent as the caring of herds among the pastoralists of the Horn. Nor is the former as closely related to identity and personal values as the latter. Taking care of teeming life by maintaining the reproductive capacity of the herds is usually seen as a more tricky business, and more a matter of direct human responsibility, than 'looking after grass', which falls more within divine jurisdiction. To the extent that God's benevolence reflects human behaviour, as the Borana of Ethiopia and Kenya believe for example, the making of peace and consensus is considered an essential part of what people can do to enhance natural reproduction. This could be regarded as 'environmental cooperation'.

Many pastoral societies also act directly to protect the environment – for example, by reserving particular pastures for certain animal species, or imposing sanctions against such practices as cutting live trees. Normally, measures against mismanagement of environmental resources should be distinguished from active investment in environmental protection. Sandford (1983, p. 97) lists six broad sets of activities that can be used to improve range productivity in pastoral areas:

- mechanical or physical work on soil, vegetation, or structures;
- planting, seeding, or re-seeding with selective species and varieties;
- burning the vegetation;
- application of chemicals;
- altering the timing, length and succession of use by livestock of particular pieces of land; and
- regulating grazing pressure in terms of numbers, species and movements of animals.

Many groups set aside certain parts of the pasture for dry-season grazing, or for the feeding of mother cows and their calves. Other activities are less common, except the maintenance of wells and the occasional burning of grass. The latter practice is debatable. The pastoralists see it as regenerative, but Western observers normally consider it extremely harmful. Depending on one's perspective, joint action in burning grass could be regarded either as 'environmental protection' or the opposite.

Social sanctions reinforcing pastoral attitudes of environmental protectionism usually work informally with the local community, or by reference to a tribal system of jurisdiction, such as Borana law. The local neighbourhood group, shaped by traditional links of ritual and practical cooperation, takes decisions about formal sanctions. During the colonial period and afterwards, such groups have sometimes been integrated in the state structure, thus becoming administrative units. However, the scope of cooperation in technical improvement of pastures or grazing patterns has remained relatively narrow and localized.

What if 'environmental cooperation' is defined more generally in terms of response to environmental crises? Ethnic or regional groups rarely undertake collectively the role of social actors. Traditional pastoralist ethnic groups were not formalized entities with unambiguous and exclusive membership, nor did they have fixed hierarchies of authority. Rather, they were entities with vague boundaries, led by elders whose authority rested mainly on kinship relations and personal status. While they had ritual authority – to curse or to bless – they were often little more than first among equals. At the apex of the political hierarchy, they were not concerned with economic matters, but with the ritual and ideological legitimation of structures for settling issues of conflict and peace. Elders enjoyed respect, but they were not expected to meddle in how the herders minded their livestock. Concerning husbandry, migration and other matters with ecological implications, each senior herd-owner acted independently. Thus, usually it was not the tribal leaders who sought contact across the borders of regional or ethnic entities, but men representing smaller units, such as the household or co-herding groups. Cooperation in times of crisis – in contrast to conflict avoidance – would in most cases take the simple form of individual families getting permission from allies to graze stock on their territory, or consist of the simple passive acceptance of strangers. Such forms of everyday cooperation remain unnoticed. Environmental conflicts are clearly visible; environmental cooperation in its simpler forms is much less evident.

To understand who cooperates with whom and at what level, and also why it is difficult to find examples of large-scale cooperation, we need to recognize certain characteristics of pastoral land rights in the Horn of Africa. The perception of land as the property of humans varies widely. Land may be seen as a bounded, clearly delimited area to which a group has inalienable rights, or as a vague space around a place of ritual, a territory of political or economic interest, or simply the place which groups happen to occupy at a particular time. In most cases, documentary validation of such rights is lacking, and they are subject to revision, depending upon what the occupiers and their neighbours accept as valid. Claims to land usually are maintained by some kind of collective action, and with reference to a collective identity. This does not mean that the use of land is collectively regulated at tribal level. On the contrary, emphasis is placed on autonomy and independence in daily decision-making, each senior herd-owner deciding for himself, or with his chosen partners, where to graze his stock.

Changes of tribal boundaries were often the aggregate result of many small movements at the household level. Migrations of large groups also occurred, usually associated with expansion into enemy territory. Such movements took place when a pastoral group had increased its wealth while its neighbours were in a weak position. Pastoralists experiencing drought or disease are not likely to start a war: they are too busy just trying to survive and save the remnants of their herds. By contrast, pastoralists who have been displaced and kept their herds, or those who have expanded their possessions, may start a war in order to safeguard their fodder supply.

The Borana Oromo of southern Ethiopia and northern Kenya do not think they own the area 'shared' by all Borana: God owns it. They have a ritual centre around the deep wells just north of the Ethiopia–Kenya border, which they define as traditional Borana land. They have an idea of the extent of the land included in 'traditional Borana areas', that is, land to which they have a historical claim. The extent of this area depends on the range of the historical perspective. A century ago, Wajir in Kenya was part of it, but Isiolo District was not. Today, according to contemporary Borana claims, backed by colonial practice, Isiolo District is considered Borana territory, although present-day security problems seem to have alienated the Borana from these grazing grounds. By tradition, Borana clans live spread all over Borana land, and the internal divisions of the Borana community are not linked to rights over particular locations. Agnatic clans have ultimate rights to certain wells, but these wells are intermingled and used also by real and classificatory in-laws of their owners. A Borana who wants to move from Ethiopia to Waso in Kenya, or vice versa, will join clansmen there, and will be able to find them almost anywhere in Borana territory. On the same principle of ethnic kinship, assistance could be mobilized with grazing and, to some extent, with new breeding stock.

There have been other pastoral communities in East Africa and the Horn where, in contrast to the Borana, a link between a group and a particular territory was recognized. Usually, this did not mean that grazing for a herd-owner was limited to the particular territory associated with his group. Turkana clans, like Borana clans, are dispersed, but the Turkana are also divided into territorial units. Storas (1990, p. 138f) writes:

> The Turkana are divided into sections (*ekitela*), subsections and other groups on different levels of scale. These units, on all levels, are attached to territories. Individuals may refer to a location as their home area (*akwap*) and a particular territory as their *ere*. The semi arid conditions in Turkana District force the pastoralists to move outside their home areas for periods of the year . . . They then may have to trespass and temporarily stay in the *ere* of others. In order to be allowed to do so, or to avoid problems, they have to establish and maintain relationships to those who claim 'ownership' of the *ere*.

The exchange of gifts, livestock, or women, created alliances between neighbouring herdsmen, partnerships which, as Sobania writes (1990, p. 5), were made possible by

> the similarity, as much as the diversity of their socio-economic adaptations to particular environmental zones . . . Inter-societal bond partnerships grew out of the mutual economic interest two individuals previously shared as trading partners, out of the sharing of a grazing area which brought alien herdsmen into prolonged contact, or out of the hospitality extended to a neighbouring traveller on a visit . . . Tradition asserts that across the various trading relationships that existed in the region [North Kenya] commodities were 'freely' given to friends as often as individuals bartered for what they were given.

Exchange networks developed during normal times could be utilized in crises, when one community experienced disease, drought or raids. In cases where ethnic groups fought the people with whom they had links of friendship, people with affines in the enemy groups often intervened as mediators in such conflicts.

Or we can take the case of the Beja in northeastern Sudan. They have specific ideas about land ownership, with specific territories linked to their agnatic clans. Ownership of land is related to the symbolic value of being hosts (extremely important to the Beja), rather than to the possibility of excluding somebody from grazing. Grazing concessions are not negotiated at tribal level, nor given in a bargain of strict proportionality; they are extended on the basis of generalized reciprocity. Anybody can graze stock anywhere, as long as he clearly recognizes the owners. He need not be a friend. On the contrary, 'the stranger is the best guest'. Such hospitality is not limited among Beja groups, but applies to members of other ethnic groups as well.

Friendship links of the type described above also linked pastoralists and cultivators. Sobania (1990, p. 7) cites what happened in the 1890s. This was the time of a major disaster in pastoral Africa, caused by the introduction of diseased livestock at Massawa. The epidemics of rinderpest and cholera that ensued were followed by drought, causing one of the major tragedies in the

history of mankind. The Samburu and Rendille of present day northern
Kenya took refuge with the Dassanetch of Ethiopia. The latter had land for
millet cultivation and allowed the Samburu and Rendille pastoralists to plant
it. They also provided them with some animals. Sobania's argument is that in
the present century the scope of migration has been limited, and so have the
safety networks. Moreover, cultivators today do not depend on their pastoralist neighbours as a market for their grain, any more than the latter need
to exchange their livestock with the cultivators. The impersonal market is less
conducive to personalized assistance than was the bartering of old times (see
Sobania, 1988, p. 229ff; 1990, p. 11ff).

Flexibility with respect to leadership patterns and land rights was emphasized above. This must be taken into account if traditional forms of local or
regional cooperation are to be integrated into modern systems in institutionalized form. The traditional pastoral system involves a constant
re-combination of available resources of livestock, capital, labour, food, etc.
The pastoral household is itself a flexible unit, attracting dependants or relo-
cating its own members to other households, according to the availability of
food. The head of the unit includes the livestock of the dependants in his herd
and also uses their labour. Whether formally 'owned' or not, most of the
larger territory the herd operates in can be exploited only through the medi-
ation of a wider social unit. This unit is of some permanence, yet part of an
ongoing process of fission and fusion, creating temporary alignments,
alliances and groupings. Ethnic and kinship units are partly 'imagined communities' (Anderson, 1983), rather than groups interacting physically.

A pastoral group is not necessarily a bounded, exclusive group. Rather, it
consists of a number of people who identify, more or less, with the idea of a
community, joined together by shared values and customs. Over the centuries, pastoral survival has depended upon mutual assistance within the
corporative ideological constructs that such communities represent. To the
extent that people in arid areas have formed any ideas of collective responsibility towards the environment, they tend to be linked to ideas of shared
identity and solidarity, but not on a global scale. Modern Western environ-
mentalism is clearly associated with global solidarity on the one hand, and on
the other with an individualism that many pastoralists consider anti-social
and inimical to their way of life.

The difficulty of delineating exclusive population groups and matching
them with land resources creates a legal problem today. What was the very
strength of the pastoral system, its flexibility and the opportunity for individuals to seek alignment with groups inside or outside the boundary of their
group, today has become its weakness because it allows outsiders to encroach
on pastoral land.

There is often hope that working through traditional institutions will yield
increased legitimacy and efficiency to development efforts. Ignoring traditional institutions often reflects counter-productive disrespect for indigenous
knowledge and competence. In the context of environmental cooperation,
there does not seem to be much scope for coopting traditional institutions

into state administrative structures, except for purposes of mobilizing ideological commitment: or, perhaps, retaining valuable pastoral knowledge,
because traditional institutions are by nature and necessity very fluid. It is
doubtful that the flexibility of traditional institutions can be wedded to the
rigidities of bureaucracy, unless fundamental internal cultural innovation
takes place. Yet, it is to be hoped that traditional forms of cooperation will
survive in the interstices of state-imposed arrangements. Closer consideration
of traditional forms of cooperation in good and bad times may draw our
attention to some of the problems that pastoral societies have to deal with,
and which have to be solved. Finally, it may give some direction to regional
cooperation at a higher level, whether this is organized on the initiative of
governments, or because of political pressure from the pastoralists themselves.

The understanding of technical and organizational constraints of traditional pastoralism as a subsistence activity has improved greatly during the
past two decades, but with relatively little impact on policy-making. Pastoral
sustainability in arid lands presupposes the ability to invest the necessary
labour in the herds at all seasons. As an illustration of the changes that interfere with environmental sustainability take, for example, the effect of the
absence of young men recruited for military service or seeking work far away
due to poverty. Their absence results in decreased mobility, leading to heavy
wear on local grazing. The absence of schoolchildren and youths from pastoral practice causes a deterioration in pastoral competence not compensated
by formal education. Increased emphasis on production for the market shifts
the strategy towards small stock, which wear down the ground-cover more
than cattle and camels.

Armed conflicts have influenced labour patterns and adversely affected
subsistence activities all over arid Northeast Africa. The loss of land due to
military activity is an aggravating factor in a situation characterized by per-
vasive loss of pastureland to agricultural expansion, commercial ranching
and game reserves. Land-use patterns are rapidly changing, and the areas
open to traditional livestock rearing continue to shrink. This problem may be
of greater importance to pastoralists in East Africa than range degradation,
though the two are closely linked.

The need for a regional approach to the problem of pastoral land loss has
long been recognized, but perhaps because of methodological difficulties and
political expediency alike, there is still a dearth of information about such
issues as changes in population pressure and migration patterns within wider
pastoral regions. Even within one country, there are usually so many agents
involved in changing land-use patterns that it is impossible for anyone to get
a clear picture of the overall situation. Regional cooperation is necessary in
order to chart all threats to pastoral sustainability through land loss, whatever
their origin. Such a venture would involve a classification of types of land
according to usefulness in different pastoral regimes. It would have to reflect
a concern for the regionally available sources of extra fodder in critical seasons, and monitor interrelations of population movements from one area of

high pressure to another. It would involve also the promotion of legal protection of rights to particular plots of land that are vital to the production system as a whole. For both those who plan the future of pastoral nomads, and the people on the range themselves – who may only see the shrinkage of resources in terms of pressure from neighbours or immigrants to their area – such a mapping might be revealing and, most probably, shocking. Let us hope that governments of the countries involved will not put obstacles in the way of attempts to achieve a holistic picture of the precarious conditions endured by a part of their population.

8.4 The Nile Basin

Given the central role of the Nile River system, a water management scheme will have to include the countries that share the Nile basin and have a vital interest in its water. The various riparian states are not similarly or equally interested. Rwanda and Burundi are marginally involved, simply because part of their territory falls in the basin of the Kagera River, Lake Victoria's largest contributor. Congo (Zaire) is similarly involved because it has a share of Lake Victoria's shore and is a riparian of the Semliki River which empties into Lake Mobutu (Lake Albert). None of these three states is dependent upon the Nile or the Equatorial Lakes for their economic welfare. Uganda's water needs are met through precipitation, and its interests centre on the production of hydroelectric power. Kenya and Tanzania are primarily interested in Lake Victoria as a source of water for irrigation, and secondarily in what they can gain from the users of water downstream in exchange for their cooperation in water management. The states vitally concerned are Egypt and the Sudan, the main consumers of water, and Ethiopia, the main supplier. Because any kind of unilateral action concerning the Nile water will prove mutually harmful, these three states have the greatest common interest in reaching agreement on a cooperative arrangement for joint water management.

Their specific interests diverge. For example, Sudan has more to gain from developing the Blue Nile–Atbara system rather than the White Nile, because the former offers gravity-flow irrigation without the need for pumping, as well as sites for hydroelectric generation, while the latter is located in a perennially insecure region. All the same, Sudan and Egypt are forced by their interdependence to cooperate, and they have a history of doing so. Ethiopia is the key to any solution based on a cooperative regional arrangement. Ignored when decisions were made in the past concerning the use of the Nile's water, Ethiopia has a historical grievance, as well as lack of experience in negotiating effectively in this field. Consequently, it has remained aloof from consultations intended to promote a regional approach. Its position, often reiterated, is based on the so-called Harmon doctrine of international law, which recognizes the absolute sovereignty of states over international rivers within their borders, in the absence of any international agreement. A contrasting position is taken by Sudan and Egypt, which invoke international customary law that requires states not to interfere with the flow of international rivers in a way that will have adverse consequences for downstream countries. This appears to be the guiding spirit also of the *Principles of Conduct in the Field of the Environment for the Guidance of States* . . ., drafted by the Governing Council of UNEP in May 1978.

Ethiopia's position is shaped by the fact that it has yet no specific plans for water utilization on a large scale, nor the capacity to formulate them. Waterbury (1987, p. 103) notes that 'hydrological expertise is very unequally distributed. Until that inequality is overcome, the least expert will have good reason to avoid negotiations in which they might unknowingly trade away some important interests.' Keenly aware of this danger, Ethiopia has rebuffed attempts to draw it into negotiations by insisting on the right of sovereignty and the principle of 'equitable sharing' of water among the riparian states, as a precondition for negotiations. Egypt, on the other hand, is anxious to secure international agreements that would safeguard its 'natural and historical' rights over the Nile. The Egyptians favour wide-ranging regional schemes that downplay the importance of water sharing, but highlight the benefits other riparian states stand to gain from integrated development projects, ranging from environmental security to tourism. There is suspicion that Egypt's real motive is to increase its own share of water, while the rest are still uncertain of their goals. Helping them to define such goals could be a major step towards cooperation on the regional level. Ethiopia could more easily be persuaded to enter regional agreements if its own objectives for water utilization were defined, and it could proceed to negotiate on that basis. It could then become clear what benefit Ethiopia might derive from a regional water management scheme. Ideally, a regional arrangement would include other international rivers in the region – those crossing from Ethiopia into Somalia, from Eritrea into Sudan, etc.

8.5 Food Security

In analysing the general failure to promote agricultural development in the Horn, numerous features of the region are commonly cited. The inadequacy of its resources, the vagaries of its climate, the ignorance of its people, rapid population increase, ecological degradation, traditional land tenure systems, archaic technologies, faulty price systems, faulty tax structures and the malignant effect of the global market are among them. Alternatively, wholesale blame is placed on the particular economic model used. Several of these – capitalist, socialist and variations of each – have been tried in all the countries of the Horn, at one time or another. Finally, the finger is pointed at the regimes ruling these countries, for not properly implementing the development strategies they have adopted.

Implicit in this analysis is the certainty that failure is one of implementation, not of conception. All development strategies are of foreign origin, introduced into Africa with scarcely any modification. Even those critics who reject one model would have it replaced by another of foreign origin. In fact, the problem is not one of implementation, but of misconception. The region's climate, backwardness, etc. are indeed real obstacles to development. However, these are precisely the factors that one would expect an appropriate strategy of development to take into account and propose to deal with. In fact, the real failure is one of conception, and is inherent in the models imported from abroad.

The common denominator of all imported development strategies is industrialization. The pursuit of this chimera imposed a set of priorities that represent a gross distortion of reality. Investment and technological innovation were concentrated in the urban sector, which houses and supports only a small proportion of a population that is overwhelmingly rural. As a result, this sector acquired a near monopoly of material and social resources as well as political power. Investment and technological innovation were brought also to agriculture, but only in order to earn the foreign exchange needed to develop the urban sector. Production of cash crops for export was promoted at the expense of food production for domestic consumption. Land, water and labour were diverted from domestic food production to production for export. Almost nothing was done for the improvement of the subsistence sector, on which the vast majority of the population depend for their existence. Given the high rate of population growth, this distortion led to a drop in food production per capita and to permanent food shortages. If anything has become clear from the experience of the countries in the Horn, it is that the priorities of their economic strategies must be reversed. Plans for social and economic development must have the welfare of the majority of the people as their immediate goal, and food security for all must have the highest priority.

8.6 Improving Regional Trade*

The free flow of food across borders is an objective that can be achieved. It requires recognition of the fact that no such flow can be conceived outside the laws of economics. On the other hand, not all measures proposed here are restricted to the realm of economics. Political choices are essential for an effective flow of food and are a leading element in the ten-point package outlined below.

1   In the current situation, politics must essentially be about food. The pursuit of an equitable and sustainable food supply becomes meaningful and achievable only if the peoples and nations in the region can consciously and positively resolve the problematic issues of politics.

*This section is part of Mulugeta Bezabeh's contribution on 'Free Flow of Food Across Borders in the Horn of Africa', presented at the PRIO/UNEP seminar in Sigtuna, Sweden, September 1991.

2   Except for the Djibouti franc, none of the currencies in the Horn is convertible. Because of high external imbalances in almost all countries in the
region, payments for official exports are made in convertible currencies. The
general effect has been to encourage inefficient imports at the expense of
exports. The degree of overvaluation differs from country to country, so that
desirable trade has become unprofitable, while economically inefficient trade
becomes profitable. It is generally recognized that currency convertibility or
overvaluation are not major constraints on the existing unofficial trade, since
the prices of the commodities involved find their own profitable levels because
of the high demand for them in the importing country. However, the strain on
and damage to the economy of non-convertibility and overvaluation of currencies are high, so the countries in the region need to adjust official exchange
rates to economic values in order to permit the gradual unification of official
and parallel exchange markets. In the first half of the 1990s, Ethiopia and
Kenya devalued their currency in accordance with the provisions of the
Structural Adjustment Programme they adopted.
3   Unofficial trade needs to be radically liberalized and regularized.
Governments need to face reality, legalize this trade, take advantage of it in
the short run and, in the long run, integrate it in the official trade system.
This will involve removing regulatory and bureaucratic constraints, such as
trade prohibitions, monopolies, etc. Other measures will involve adopting
realistic pricing systems, reform of tax systems relating to trade, and providing infrastructure and market information, especially in border areas. Such
measures will not only help traders, but also provide governments with tax
revenue, and enable them to collect reliable statistical information to be used
for purposeful planning. Many of the measures advocated here were adopted
in Ethiopia, Kenya and Eritrea in the early 1990s within the framework of the
Structural Adjustment Programme.
4   There is a need for substantial investment in infrastructure, particularly in
market structures which can improve the profitability of food trade by reducing the real costs of marketing. A significant unexploited trading potential
exists within and between countries in the Horn, partly explained by the
inability of marketing parastatals and private marketing to fulfil their roles
efficiently. While privatization of a large part of trade is necessary, parastatals
will still have a role to play, especially in holding food security stocks.
5   Of immediate relevance to the free flow of food within the region are the
border production areas, particularly the northwestern lowlands on both
sides of the Ethiopia–Sudan border, the Shebeli region on both sides of the
Ethiopia–Somalia border, the Ogaden, and the eastern lowland of Ethiopia
bordering Djibouti. The first two areas have a high potential for sorghum
production, and the last two for livestock. These are areas where a substantial trade in food is going on, most of it unofficial. These areas have the same
or nearly the same environment for production, and unless policies for further
development are coordinated between countries, the simple pursuit of food
self-sufficiency by each country can have a regressive influence on intra-
regional trade. Quite apart from this, income opportunities will be lost if the

countries fail to explore the possibilities of producing other crops of high value and demand in the Horn or elsewhere. Joint policy and programme coordination for the border areas is a necessary precondition for healthy development there.

6   Food aid has been given readily to all countries in the Horn. The depressive effects of an increasing food supply on national production are well known. Furthermore, food aid has influenced patterns of consumption and trade to such an extent that commercial imports are following the route of food aid commodities. Harmonization of food aid policies, similar to what has been done in the SADCC countries, may be necessary, especially if and when local production increases substantially. Countries in the Horn may need to harmonize their policies in order to evolve arrangements to raise foreign exchange, such as the Triangular Food Aid Transactions. These resources could be used for food purchases with hard currencies within the region. Local purchases of surplus food could also be considered, in consultation with local donors.

7   A well-structured information system in support of a free flow of food within the region could have strong long-term benefits. Trade, price and other statistical data related to production, consumption, storage, etc. could be assembled in regularized form and put at the disposal of the countries in the Horn. This could play a crucial role in the planning and coordination of food trade in the region.

8   Those directly involved in the formal and informal flow of food across borders have to be apprised continuously of the prevailing policies and incentives provided to facilitate trade. It is important that information should not be confined to national marketing officers, but must permeate down to the grass-roots level.

9   One critical issue that negatively affects trade, especially food trade, is the absence of credit and finance in the formal sector. Traditionally, food trade in the Horn has been a small-scale undertaking, and one rarely finds traders with operational and physical infrastructures sufficient to organize trade in a modern and sustainable way. The lack of credit resources accounts for the absence of modern methods and procedures in the food trade.

10   Should the countries in the Horn form a regional grouping for food matters, or even for all-inclusive economic matters? Or should this be limited to simple mechanisms of coordination? This is a subject that merits further investigation.

8.7 Conflict

Political life in Africa is bedevilled by the 'curse of the nation-state', as Davidson (1992) put it. The states in the Horn were fashioned on the nation-state model of Western European provenance, whose categorical imperative is cultural homogeneity. This model was adopted regardless of the ethnic and cultural diversity of this region, and was forcefully promoted under the guise of national integration. In cases like Ethiopia and Sudan, this amounted to a process of forced assimilation into the culture of the dominant nationality. Centralization, another feature of the nation-state model, ensured a monopoly of power for the dominant group. Given the commanding role of the post-colonial state in the production and distribution of material and social resources, access to state power usually meant access to these resources, and the reverse also held true.

As a result, the state became the target of those who were dissatisfied with the existing order of things and sought to change it. In view of the ethnocratic nature of the state – the monopolization of state power by one or more ethnic groups, and their assimilationist policies – the opposition found it both necessary and effective to rally popular support on ethnic grounds. Consequently, the manifold conflict over state power and resources was dubbed 'ethnic', a label uncritically applied to all kinds of social disorder in this region.

The state and the policies it promoted were not irrelevant to all the people, because obviously the interests of certain groups were well served. The political base of these groups was very narrow, however, and in order to retain exclusive control of the state they were compelled to resort to force. Thus, the state in the Horn of Africa became increasingly arbitrary and violent, and in the final stage waged genocidal wars against its own subjects. The means for such excesses were provided by foreign powers which had their own reasons for such generosity. When these reasons ceased to be valid, outside support was withdrawn, and the state in Somalia and Ethiopia collapsed, while in Sudan and Djibouti the struggle continues.

What the political experience of the states in the Horn has shown is that narrowly based, centralized and authoritarian forms of state rule will not endure without massive material support from abroad. Changes in the international scene indicate that such support is not likely to be forthcoming in the future. The restructuring of the states in the region, therefore, must recognize ethnic and regional diversity and allow for its political expression. A new type of state will have to evolve, and its design will have to conform to local conditions and the limitations they impose. Decentralization, the common demand of all the movements that fought against the post-colonial state, should be a key feature of the future state. With effective decentralization, cultural pluralism should prevail and ethnic groups should be able to assert their identities.

A process of geopolitical transformation got under way in the Horn with the independence of Eritrea and the secession of northern Somalia (Somaliland) in the early 1990s. At the same time, the struggle in Sudan took a new turn, when the Southern Sudanese resistance movement changed its objectives and made self-determination its goal, indicating that the breakup of Sudan was a likely outcome of the conflict. A similar fate may await Djibouti, given the intensifying rivalry between Somali and Afar. The fate of Ethiopia hung in the balance, as the regime that came to power in 1991 staked the future of the country on a bold experiment in statecraft. A highly decentralized federal state structure was designed, based on self-governing ethnic units, each with its own official language and education system. Ethnicity became the defining political principle in the country, and ethnic groups mobilized to defend territorial, economic, cultural and political claims. Granting political priority to ethnicity is unprecedented in post-colonial Africa, and the outcome of the Ethiopian experiment still remains uncertain. Its failure would threaten the survival of the state with disintegration. Success, however, could make it a model for other states in Africa. In Somalia, the total collapse of the post-colonial state presented an opportunity to fashion state structures suitable to a society based on the clan system and infused with the ethic of nomadic pastoralism. Crafting such structures requires a high order of ingenuity and innovation, as well as inspiration from Somali tradition.

The demise of the post-colonial state was not mourned by many in the Horn, because its record was one of dismal failure. The proliferation of smaller states is not in itself an improvement, and could mean a turn for the worse if these states cling to the antagonistic exclusiveness and inflexible sovereignty of their predecessors. This would mean increased fragmentation and a proliferation of obstacles to economic, social and political interaction for the people of the region. However, pressure on the smaller state units to cooperate, coordinate and even integrate their activities would be great, for they would be unable to accomplish much on their own.

Economic integration in the form of a regional common market would be an obvious first step. Ethiopia and Eritrea moved in that direction by not imposing barriers to the movement of goods and people in the newly formed border between them. A similar situation existed unofficially in the borders between Ethiopia and the two parts of Somalia, mainly because state controls had collapsed. The cost of defeat for Ethiopia was moderated by making Asab a free port. A similar arrangement for Massawa could prove of benefit to Sudan and Eritrea, especially if the two countries were linked by rail. Likewise, the Ogaden and southeastern Ethiopia could benefit if such an arrangement could be made with Somaliland for Hargeisa.

Lowering frontier barriers could best be accomplished on a bilateral basis. Steps may include open corridors (for pastoralists crossing, for instance), free zones where people from both sides can mingle freely, joint administration of border regions and plural citizenship. By evading the prickly issue of sovereignty, the last two could apply most beneficially to areas that are bones of contention between two states. For instance, regional autonomy for the Ogaden and dual (Ethiopian–Somali) citizenship for its Somali inhabitants would remove a historic grievance and a potent source of conflict in the region. Joint administration of the Haud pastureland could be the way to diffuse the conflict between the Ogaden and Ishaq clans in that area. Similar arrangements between Ethiopia, Eritrea and Djibouti could defuse the problem created by the partition of the Afar.

Political practice reflecting African reality may have to go against Western conventional conceptions of the nature of the state. The wedlock of nation and state may have to be dissolved, the nation-building exercise abandoned, and multi-nationalism, reflecting the reality of ethnic pluralism, accepted. Following the end of the Cold War, democracy was recommended as a panacea for all the problems of Africa, including economic stagnation. Democracy appeared suspended in a historical void, as if it had no roots in time and space, and no relation to other social processes and institutions. Scant notice was taken of the fact that democracy in practice, not in theory, is a very recent development in the history of the West; that it coincides with industrialization; and is based on a balance of social forces, as well as on a host of other inter-related socio-economic developments. Most of Africa, the Horn included, manifestly lacks the material and social foundations for Western-style representative democracy. If Africans are to participate meaningfully in making decisions that affect them, which is true democracy, rather than simply electing others to make the decisions for them, they will have to fashion their own ways of doing so, as indeed they have done in the past.

Given political stability and broad popular participation in government, the countries of the Horn ought to be able to pursue a modest development agenda, also in conformity with local conditions and the limitations they impose. These limitations, described and analysed in detail in the preceding chapters, rule out dramatic change in the material condition of the people of the region in the foreseeable future. The most that can be hoped for is to prevent further deterioration, and even that would require an enormous effort on the part of the people themselves.

The West must help, not simply offering charitable support, but also by removing some of the crippling economic burdens it has imposed on developing countries. These burdens have included debt servicing, world market price structures that steadily reduce the value of primary exports, tariff barriers that impede the development of manufacturing capacity in developing countries, the high price of technological innovation, etc. Moreover, the developed world must re-evaluate its commitment to the post-colonial state, and allow for the possibility that flexibility and diversity in modes of government fashioned by Africans themselves might well be more suitable to the continent.

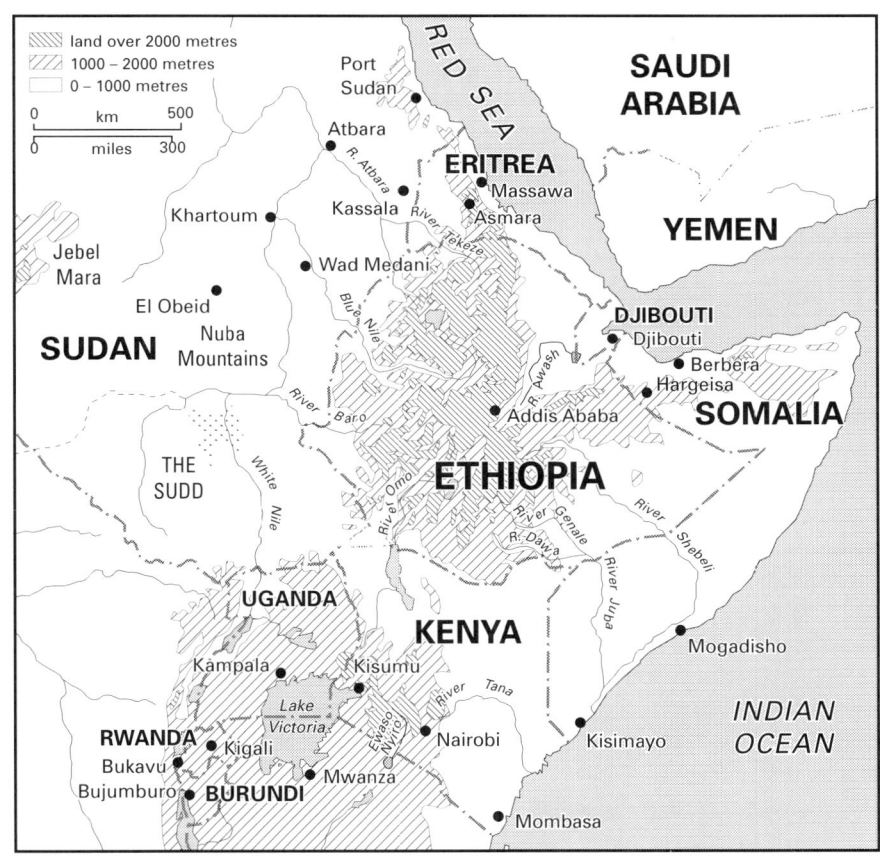

Map 1 *Physical Map of the Horn of Africa*

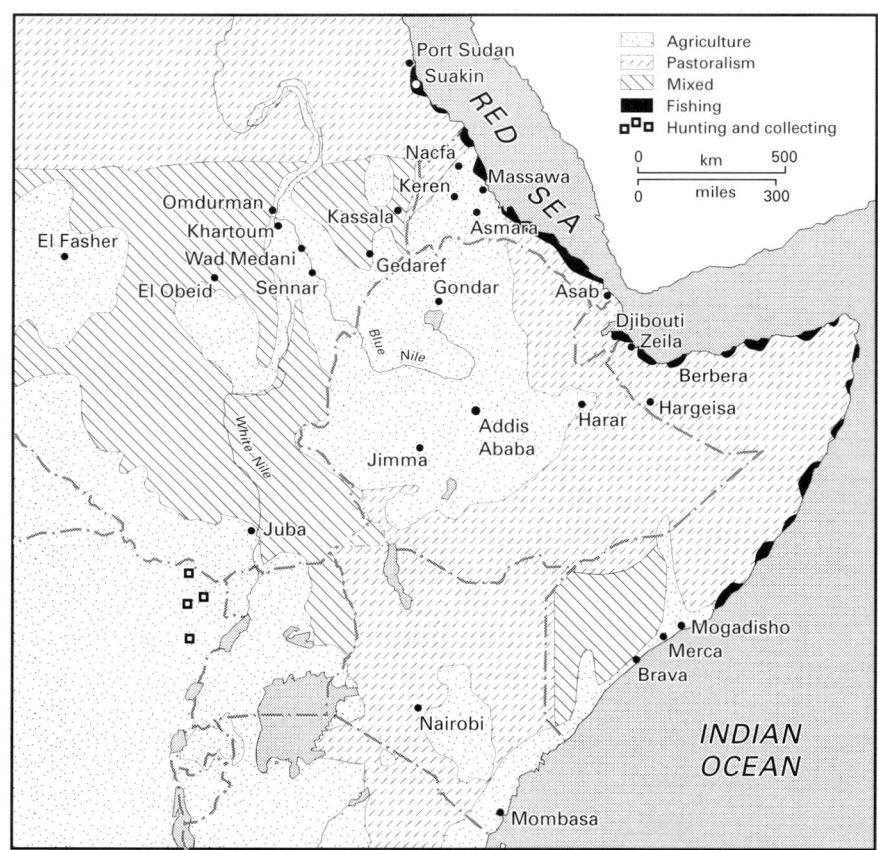

Map 2   *Modes of Production in the Horn of Africa*

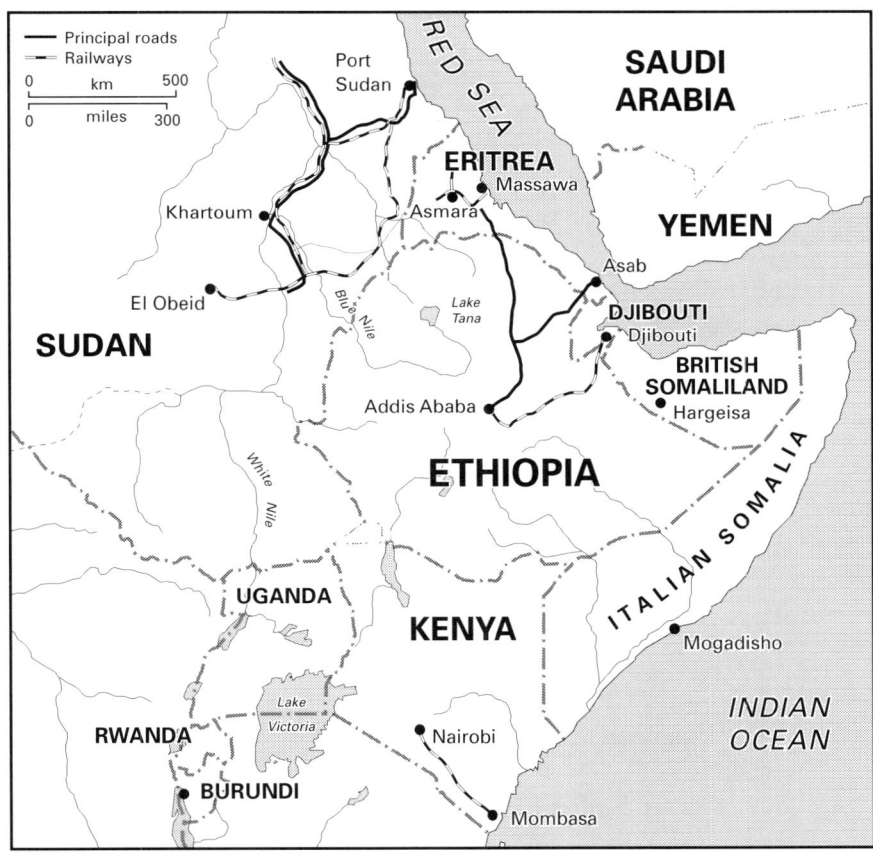

Map 3   *Colonial Map of the Horn of Africa*

Map 4   *The States of the Horn of Africa*

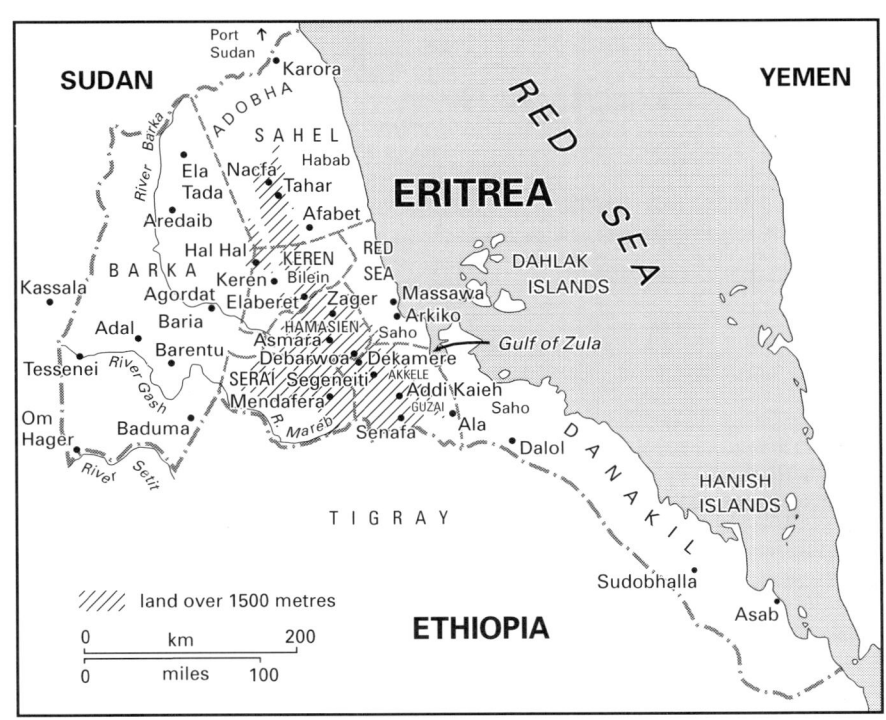

Map 5   *Eritrea*

Map 6   *Southern Sudan*

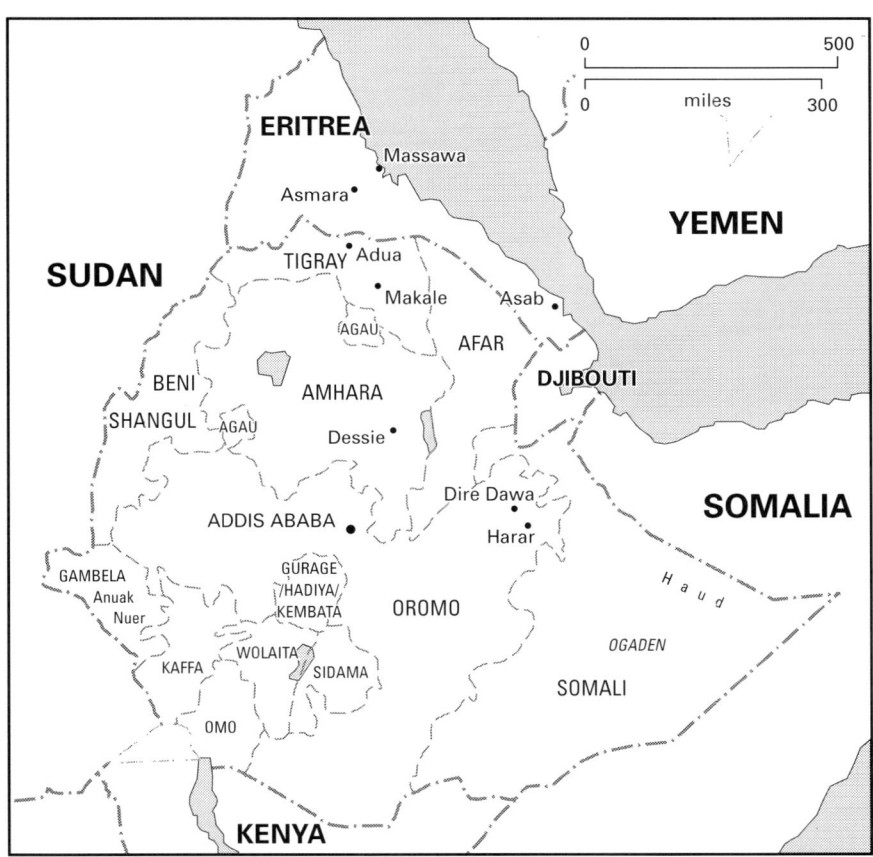

Map 7   *Ethnol Regional Administrative Divisions of Ethiopia*

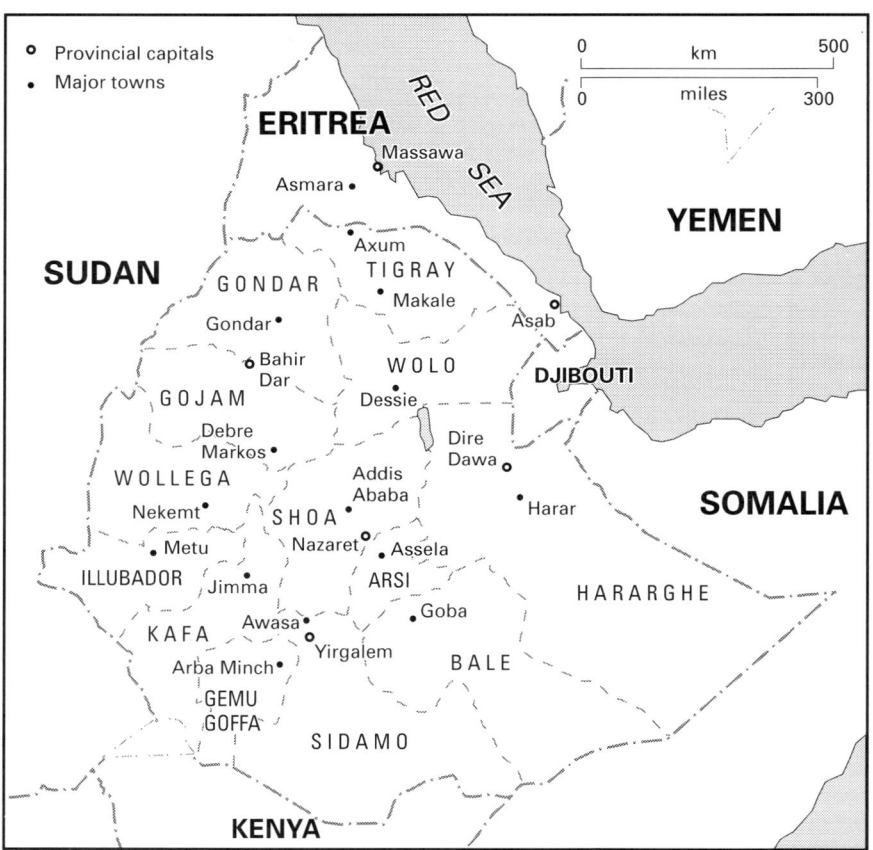

Map 8   *Former (pre-1991) Regional Divisions of Ethiopia*

References

Abdel Ghaffar, M.A., ed., 1976. *Some Aspects of Pastoral Nomadism in the Sudan.* Khartoum: Sudan National Population Committee.

Abdel Ghaffar, M.A., 1982. 'Planning and the Neglect of the Pastoral Nomads in the Sudan', in Gunnar Haaland, ed., *Problems of Savannah Development, Sudan Case.* Bergen: Bergen University, Occasional Papers in Social Anthropology, No. 19.

Abdel Ghaffar, M.A., 1987. 'National Ambivalence and External Hegemony: The Negligence of Pastoral Nomads in the Sudan', pp. 129–148 in M. A. Mohamed Salih, ed., *Agrarian Change in the Central Rainlands: Sudan.* Uppsala: Scandinavian Institute of African Studies.

Abdi Ismail Samatar, 1989. *The State and Rural Transformation in Northern Somalia.* Madison, WI: University of Wisconsin Press.

Adam, Jonathan S. & Thomas O. McShane, 1992. *The Myth of Wild Africa: Conservation Without Illusion.* New York: Norton.

Africa Watch, 1992. 'Sudan: Refugees in Their Own Country', vol. 4, no. 10, July.

African Academy of Sciences, 1987. *Soil and Water Management and Biotechnology in Africa.* Nairobi: AAS.

African Rights, 1993a. *The Nightmare Continues: Abuses Against Somali Refugees in Kenya.* London: African Rights.

African Rights, 1993b. *Land Tenure, the Creation of Famine, and Prospects for Peace in Somalia.* London: African Rights.

Ali Goubba, 1993. *Djibouti: Une Nation en Otage.* Paris: L'Harmattan.

Alier, Abel, 1990. *Southern Sudan: Too Many Agreements Dishonoured.* Exeter: Ithaca Press.

Anderson, Benedict, 1983. *Imagined Communities. Reflections on the Origin and Spread of Nationalism.* London: Verso.

Bennett, Olivia, ed., 1991. *Greenwar: Environment and Conflict.* London: Panos.

Berhane Woldegabriel, 1991. 'Rule by the Gun', pp. 104–106 in *Greenwar: Environment and Conflict.* London: Panos.

Besteman, Catherine, 1994. 'Individualisation and the Assault on Customary Tenure in Africa: Title Registration Programmes and the Case of Somalia', *Africa*, vol. 64, no. 4, pp. 485–503.

Bondestam, Lars, 1974. 'Peoples and Capitalism in the Northeast Lowlands of Ethiopia', *Journal of Modern African Studies*, vol. 12, no. 9, pp. 423–439.

Bonfiglioli, Angelo M., 1992. *Pastoralists at a Crossroads.* Nairobi: Project for Nomadic Pastoralists in Africa.

Boserup, Ester, 1981. *Population and Technological Change.* Chicago, IL: University of Chicago Press.

Boserup, Ester, 1993. *The Conditions of Agricultural Growth.* London: Earthscan.

Burr, Millard, 1993. *Quantifying Genocide in the Southern Sudan, 1983–1993.* Washington, DC: United States Commission for Refugees.

Butzer, Karl W., 1961. 'Climatic Changes in Arid Regimes since the Pliocene', pp. 31–56 in *Arid Zones Research*, vol. 17.

Butzer, Karl W., 1971. *Recent History of an Ethiopian Delta.* University of Chicago, Geography Department, Research Paper 136.

Cassens, Robert, 1994. *Population and Development: Old Debates, New Conclusions.* Washington, DC: Overseas Development Council.

Christianson, Carl, 1991. 'Use and Impacts of Chemical Pesticides in Smallholder Agriculture in the Central Kenya Highlands', in Carl Folke & Tomas Kaberger, eds, *Linking the National Environment and the Economy: Essays from the Eco-Eco Group*. Dordrecht: Kluwer.

Crisp, Jeff & Nick Cater, 1990. 'The Human Consequences of Conflict in the Horn of Africa: Refugees, Asylum and the Politics of Assistance', paper presented to the Regional Security Conference organized by the Institute of Strategic Studies. Cairo, 27–30 May.

Dahl, Gudrun, 1991. 'The Beja of Sudan and the Famine of 1984–1986', *Ambio*, vol. 20, no. 5, pp. 189–191.

Davidson, Basil, 1992. *The Black Man's Burden: Africa and the Curse of the Nation-State.* London: Currey.

de Waal, Alex, 1990. 'A Re-assessment of Entitlement Theory in the Light of the Recent Famines in Africa', *Development and Change*, vol. 21, no. 3, pp. 469–490.

de Waal, Alex, 1993. 'Starving out the South', pp. 157–185 in M.W. Daly & Ahmad Alawad Sikainga, eds, *Civil War in the Sudan*. London: British Academic Press.

DECARP (Desert Encroachment Control and Rehabilitation Programme), 1976. *Report*. Khartoum: National Council for Research.

Deudney, Daniel, 1990. 'The Case Against Linking Environmental Degradation and National Security', *Millennium*, vol. 19, no. 3, pp. 461–476.

Deudney, Daniel & Richard Mathew, eds, forthcoming. *Contested Ground: Security and Conflict in the New Environmental Politics*. Albany, NY: State University of New York.

Djibouti, 1990–91. *Bulletin Semestriel de Statistique*, no. 43–44. Djibouti: Ministries of Commerce, Transport and Tourism.

Dokken, Karin & Nina Graeger, 1995. *The Concept of Environmental Security – Political Slogan or Analytical Tool.* PRIO Report 2/95. Oslo: International Peace Research Institute.

Doornbos, Martin & John Markakis, 1994. 'Society and State in Crisis: What Went Wrong in Somalia?', pp. 12–18 in M.A. Mohamed Salih & Lennart Wohlgemuth, eds, *Crisis Management and the Politics of Reconciliation in Somalia*. Uppsala: Scandinavian Institute of African Studies.

Dreze, Jean & Amartya K. Sen, eds, 1990. *The Political Economy of Hunger*, 3 vols. Oxford: Clarendon Press.

Earthscan, 1984. 'Environment and Conflict: Links Between Ecological Decay, Environmental Bankruptcy and Political and Military Instability'. London: Briefing Document No. 40. Prepared by Lloyd Timberlake and Jon Tinker, also published in the *Socialist Review* (1985). See listing below under their names.

Ehrlich, Paul R., 1990. *The Population Explosion.* New York: Simon & Schuster.

El Tay et al., 1988. 'The Population of Sudan. Implications of Current Trends', in Atif A. Rahman Saghayroun et al., eds, *Population and Development: Proceedings of the Third National Population Conference, October 10–14, Khartoum*. Chapel Hill, NC: University of North Carolina at Chapel Hill.

El Tigani, ed., 1995. *War and Drought in Sudan: Essays in Population Displacement.* Orlando, FL: University of Florida Press.

EPLF (Eritrean Peoples Liberation Front)/ERA (Eritrean Relief Association), 1987. *Eritrea Food and Agriculture Assessment Study*. Final Report.

Eritrean Provisional Government, 1993. *Basic Facts.* Asmara.

Ethiopia, 1973. *Final Report of the Crop Conditions Survey for the 1972–73 Harvest*. Addis Ababa: Ministry of Agriculture.

Ethiopian Transitional Government, 1993. *Statistical Abstract, 1992*. Addis Ababa: Central Statistical Office.

Evans-Pritchard, Edward E., 1940. *The Nuer*. Oxford: Clarendon Press.

Falkenmark, Malin, 1987. 'Water-Related Constraints to African Agriculture in the Next Few Decades', pp. 439–453 in *Water for the Future: Hydrology in Perspective*. Velp: International Association of Hydrological Studies (IAHS), Publication No. 164.

Falkenmark, Malin, 1991. 'Rapid Population Growth and Water Scarcity: The Predicament in Tomorrow's Africa', pp. 81–94 in K. Davies & M. Bernstein, eds, *Resources, Environment and Population.* Washington DC: World Population Council.

Falkenmark, Malin; Jan Lundqvist & Carl Widstrand, 1989. 'Macro-scale Water Scarcity

Requires Micro-scale Approaches: Aspects of Vulnerability in Semi-arid Development', *Natural Resources Forum*, vol. 13, no. 4, pp. 258–267.

FAO (Food and Agriculture Organization), 1949. Report of the Near East Pre-Conference Regional Meeting, Beirut. FAO document C.49/1/6. Rome: FAO.

FAO (Food and Agriculture Organization), 1984. *Ethiopian Highlands Reclamation Study*. Addis Ababa: FAO.

FAO (Food and Agriculture Organization), 1984. *Production Yearbook 1983*, vol. 37. Rome: FAO.

FAO (Food and Agriculture Organization), 1985. *Production Yearbook 1984*, vol. 38. Rome: FAO.

FAO (Food and Agriculture Organization), 1986a. *Ethiopian Highland Reclamation Study*. Rome: FAO.

FAO (Food and Agriculture Organization), 1986b. *African Agriculture. The Next 25 Years*. Annex 3, Rome: FAO.

FAO (Food and Agriculture Organization), 1987. *Crop Assessment Mission, Eritrea/Ethiopia*. Rome: FAO.

FAO (Food and Agriculture Organization), 1987. *Production Yearbook 1986*, vol. 40. Rome: FAO.

FAO (Food and Agriculture Organization), 1988. *Production Yearbook 1987*, vol. 41. Rome: FAO.

FAO (Food and Agriculture Organization), 1991. *Food Aid in Figures*. Vol. 8/1. Rome: FAO.

FAO (Food and Agriculture Organization), 1991a. *Production Yearbook*. Rome: FAO.

FAO (Food and Agriculture Organization), 1991b. *The State of Food and Agriculture*. Rome: FAO.

FAO (Food and Agriculture Organization), 1992. *Production Yearbook 1991*, vol. 45. Rome: FAO.

FAO (Food and Agriculture Organization), 1993a. *Ethiopia: Agricultural Sector Review*. Rome: FAO.

FAO (Food and Agriculture Organization), 1993b. Special Alert No. 243, 17 December 1993. Rome: FAO.

FAO (Food and Agriculture Organization), 1993c. *The State of Food and Agriculture*. Rome: FAO.

FAO (Food and Agriculture Organization), 1995. *Production Yearbook 1994*, vol. 48. Rome: FAO.

Firebrace, James & Gayle Smith, 1982. *The Hidden Revolution: an Analysis of Social Change in Tigray, Northern Ethiopia*. London: War on Want.

Fouad Ibrahim, N., 1984. *Ecological Imbalance in the Republic of Sudan with Reference to Desertification in Darfur*. Bayreuth: Bayreuther Geowissenschaftliche Arbeiten.

Gaim Kibreab, 1987. *Refugees and Development in Africa*. Trenton, NJ: Red Sea Press.

Garang, John, 1987. *John Garang Speaks*. London: Kegan Paul.

Geshekter, Charles L., 1982. 'British Imperialism in the Horn of Africa and the Somali Response, 1884–1899', PhD. thesis, University of California at Los Angeles.

Gillis, John R., Louise A. Tilly & David Levine, eds, 1992. *The European Experience of Declining Fertility: The Quiet Revolution*. Oxford: Blackwell.

Graeger, Nina & Dan Smith, eds, 1994. *Environment, Poverty, Conflict*. PRIO Report 2/94. Oslo: International Peace Research Institute.

Graham, Olivia, 1989. 'A Land Divided: The Impact of Ranching on a Pastoral Society', *The Ecologist*, vol. 19, no. 5, pp. 180–185.

Griffiths, John F., 1972, 'The Horn of Africa' (ch. 4) and 'The Ethiopian Highlands' (ch. 5), in John F. Griffiths, ed., *Climates of Africa*. Amsterdam: Elsevier.

Grimstad, Per O., 1991. 'A Future for Development'. Keynote Address, International Symposium on Technology on Sustainable Development in Sub-Saharan Africa. Stavanger: NORAD.

Guariso, Giorgio & Dale Whittington, 1987. 'Implications of Ethiopian Water Development for Egypt and the Sudan', *Water Resources Development*, vol. 3. no. 2, pp. 105–114.

Hawando, Tamirie, 1994. *Institutional Capacity for Assessing Land Degradation in Eritrea*. Final Report. Nairobi: United Nations Environment Programme.

Head, Stephen M., 1987. 'Red Sea Fisheries', in A.J. Edwards & Stephen M. Head, *Red Sea*. Oxford: Oxford University Press.

Hellden, Uit, 1984. 'Drought Impact Monitoring by Remote Sensing Studies of Desertification in Kordofan'. Lund: Lund University, Geography Department, *Rapporter & Notiser*, 61.

Hemming, Charles F., 1966. 'The Vegetation of the Northern Region of the Somali Republic', *Proceedings of the Linnean Society*, vol. 177, no. 2, pp. 173–250.

Heyer, Judith, 1990. 'Poverty and Food Deprivation in Kenya's Smallholder Agricultural Areas', pp. 236–280. in Jean Dreze & Amartya K. Sen, eds, *The Political Economy of Hunger*, vol. 3. Oxford: Clarendon Press.

Hodge, Carl & Peter C. Duisberg, eds, 1963. *Aridity and Man*. Washington, DC: American Association for the Advancement of Science.

Hogg, Richard, 1988. 'Water Harvesting and Agricultural Production in Semi-arid Kenya', *Development and Change*, vol. 19, no. 1, pp. 69–87.

Holt, Julius F., J. Seaman & J.P. Rivers, 1975. 'The Ethiopian Famine of 1973–74', *Proceedings of the Nutrition Society*, vol. 34, pp. 115–116.

Homer-Dixon, Thomas F., 1994. 'Environmental Scarcities and Violent Conflict: Evidence from Cases', *International Security*, vol. 19, no. 1, pp. 3–40.

Hultin, Jan, 1988. *Farmers' Participation in Wollo Programme*. Department of Social Anthropology, University of Stockholm.

Human Rights Watch/Africa, 1994. *Civilian Devastation: Abuses by All Parties in the War in Southern Sudan*. New York: Human Rights Watch.

Hunting Technical Services, 1975. *Rural Development Study in Tigre, Ethiopia*. Study published by Hunting Technical Services Ltd, Hertfordshire, UK.

Huntingford, G.W.B. (ed. and trans.), 1980. *Periplus of the Erythraean Sea*. London: Hakluyt Society.

Hurni, Hans, 1987. 'Erosion – Productivity – Conservation Systems in Ethiopia', *Proceedings of the 4th ISCO Conference*, Venezuela (University of Berne).

IGADD (Inter-Governmental Authority on Drought and Development), 1989. *Study of the Potential for Intra-Regional Trade in Cereals in the IGADD Region*. Djibouti: IGADD.

IGADD (Inter-Governmental Authority for Drought and Development), 1990. *Food Security Strategy*. Djibouti: IGADD.

ILO (International Labour Organization), 1989. *Generating Employment and Incomes in Somalia*. Addis Ababa: ILO.

IUCN (International Union for Conservation of Nature), 1989. *The IUCN Sahel Studies*. Gland, Switzerland.

IUCN (International Union for the Conservation of Nature), 1991. *Fighting for Survival: Insecurity, People and the Environment in the Horn of Africa*. Gland, Switzerland.

Kamel Mansour, 1972. 'Environment and Crop Production in the Sudan', Proceedings of the Arab Regional Symposium on Environmental Aspects and Development in Arab Countries. Khartoum, 5–12 February.

Karuti, Kanyinga, 1993. 'The Social-Political Context of the Growth of Non-governmental Organizations in Kenya', pp. 53–77 in Peter Gibbon, ed., *Social Change and Economic Reform in Africa*. Uppsala: Scandinavian Studies Institute.

Keen, David, 1992. *A Political Economy of Refugee Flows from the South-West Sudan, 1986–1988*. Geneva: United Nations Research Institute for Social Development, Discussion Paper, November.

Kenya, 1989. *Development Plan, 1989–1993*. Nairobi: Government Printer.

Kenya, 1990. *Household Food Security and Nutrition Policy*. Nairobi: Government Printer.

Kenya, 1995. *Economic Survey, 1995*. Nairobi: Central Bureau of Statistics.

Khalid, Mansour, 1984. 'The Nile Waters: The Case for an Integrated Approach', pp. 8–24 in Mohammed O. Beshir, ed., *The Nile Valley Countries. Continuity and Change*. University of Khartoum: Institute of African and Asian Studies.

Kilewe, A.M., & D.B. Thomas, 1992. 'Land Degradation in Kenya'. London: Commonwealth Secretariat.

Kraus, E.B., 1955. 'Secular Changes of Tropical Rainfall Regimes', *Quarterly Journal of the Royal Meteorological Society*, vol. 81, no. 349, pp. 430–439.

Kumar, B.G., 1990. 'Ethiopian Famines 1973–1985', pp. 173–216 in Jean Dreze & Amartya K. Sen, eds, *The Political Economy of Hunger*, vol. 2. Oxford: Clarendon Press.

Lako, George T., 1985. 'The Impact of the Jonglei Scheme on the Economy of the Dinka', *African Affairs*, vol. 84, no. 334, pp. 15–38.

Lako, George T., 1992. 'The Jonglei Canal Scheme as a Socio-economic Factor in the Civil War in the Sudan', pp. 45–58 in Michael B.K. Darkoh, ed., *African River Basins and Dryland Crises*. Uppsala: Uppsala University, Geography Department.

Larson, Barbara A. & Daniel W. Bromley, 1991. 'Natural Resource Prices, Export Policies, and Deforestation: The Case of the Sudan', *World Development*, vol. 19, no. 10, pp. 1289–1297.

Lewis, Ioan M., 1958. 'Modern Political Movements in Somaliland', *Africa*, vol. 4, no. 28, pp. 344–363.

Lewis, Ioan M., 1980. *The Modern History of Somaliland: From Nation to State*. New York: Longmans.

Little, Peter D., 1992. *The Elusive Granary*. Cambridge: Cambridge University Press.

Lodgaard, Sverre & Anders Hjort of Ornäs, eds, 1992. *The Environment and International Security*. PRIO Report 3/92. Oslo: International Peace Research Institute.

Maffi, M., 1975. *Wollo: Two Years after the Crisis*. Addis Ababa: Consolidated Food and Nutrition System.

Maknun Gamaledin, 1993. 'The Decline of Afar Pastoralism', pp. 45–62 in John Markakis, ed., *Conflict and The Decline of Pastoralism in the Horn of Africa*. London: Macmillan.

Mamdani, Mahmood, 1972. *The Myth of Population Control: Family, Caste and Class in an Indian Village*. New York: Monthly Review Press.

Markakis, John, 1974. *Ethiopia: Anatomy of a Traditional Polity*. Oxford: Clarendon Press.

Markakis, John, 1987. *National and Class Conflict in the Horn of Africa*. Cambridge: Cambridge University Press.

Markakis, John, ed., 1993. *Conflict and the Decline of Pastoralism in the Horn of Africa*. London: Macmillan.

Mazrui, Ali, 1975. *Soldiers and Kinsmen in Uganda: The Making of a Military Ethnocracy*. Beverly Hills, CA: Sage.

Mekuria Bulcha, 1988. *Flight and Integration: Causes of Mass Exodus from Ethiopia and Problems of Integration in Sudan*. Uppsala: Scandinavian Institute of African Studies.

Mesfin Wolde Mariam, 1986. *Rural Vulnerability to Famine in Ethiopia, 1958–1977*. London: Intermediate Technology Press.

Miller, Norman N., 1984. *Kenya: Quest for Prosperity*. Boulder, CO: Westview Press.

Mohamed Salih, M.A., ed., 1987. *Agrarian Change in the Central Rainlands: Sudan*. Uppsala: Scandinavian Institute of African Studies.

Mohamed Salih, M.A., 1990a. 'Pastoralism and the State in Africa: An Overview', *Nomadic Peoples*, 25–27, pp. 7–18.

Mohamed Salih, M.A., 1990b. 'Tribal Militias, the SPLA and the Sudanese State', pp. 65–82 in Abdel M. Ahmed & Gunnar Sørbø, eds, *The Management of the Crisis in Sudan*. Bergen: University of Bergen, Centre for Development Studies.

Mohamed Salih, M.A., ed., 1994. *Inducing Food Security*. Uppsala: Scandinavian Institute of African Studies.

Mohammed Hassen, 1990. *The Oromo of Ethiopia*. Cambridge: Cambridge University Press.

Mohammed Omer Beshir, 1975. *The Southern Sudan: From Conflict to Peace*. London: Hurst.

Moore Lappe, Frances & Rachel Schurman, 1988. *The Missing Piece in the Population Puzzle*. San Francisco, CA: Institute for Food and Development Policy.

Morgan, W.T.W., 1973. *East Africa*. London: Longman.

Mubarak, Jamil Abdalla, 1996. *From Bad Policy to Chaos in Somalia*. Westport, CT: Praeger.

Mustafa M. Khogali, 1991. 'Famine, Desertification and Vulnerable Populations: the Case of Umm Ruwaba District, Kordofan Region, Sudan', *Ambio*, vol. 20, no. 5, pp. 204–206.

National Academy of Science, 1994. *Population Summit of the World's Scientific Academies*. Washington, DC: The National Academy Press.

National Atlas of Ethiopia, 1988. Addis Ababa: Mapping Authority.

National Population Policy of Ethiopia, 1993. Addis Ababa: Office of the Prime Minister.

NOPA (Project for Nomadic Pastoralists in Africa), 1992. *Pastoralists at a Crossroads*. Nairobi: NOPA.

NORAD (Norwegian Agency for Development Cooperation), 1990. *NORAD in the Nineties*. Oslo: NORAD.

Ochieng, J.A.W., 1985. 'Maize Research in Kenya: An Overview', pp. 26–45 in Proceedings of the First Eastern, Central and Southern Africa Regional Maize Workshop, Lusaka, March 10–17.

ODA (British Overseas Development Administration), 1994. *Kenya: Profile of Agricultural Potential*. London.

Painchaud, Paul, ed., 1990. *Geopolitical Perspectives on Environmental Security*. Cahier du GERPE, No. 92–05. Quebec: Université Laval.

Pankhurst, Alula, 1992. *Resettlement and Famine in Ethiopia*. Manchester: Manchester University Press.

Pankhurst, Richard, 1968. *Economic History of Ethiopia*. Addis Ababa: Haile Selassie University Press.

Pankhurst, Richard, 1985. *The History of Famine and Epidemics in Ethiopia Prior to the Twentieth Century*. Addis Ababa: Relief and Rehabilitation Commission.

Panos Institute, 1993. *Grasshoppers and Locusts: The Plague of the Sahel*. London: Panos.

Parnwell, Mike, 1993. *Population Movements and the Third World*. London: Routledge.

Pearce, David W. & Kerry R. Turner, 1990. *Economics of Natural Resources and the Environment*. Baltimore, MD: Johns Hopkins University.

Peberdy, J.R., 1963. 'Rangeland', pp. 173–176 in W.T.W. Morgan, ed., *East Africa: People and Resources*. Nairobi: Oxford University Press.

Pickett, James, 1991. *Economic Development in Ethiopia*. Paris: Organization for Economic Cooperation and Development (OECD).

Rasool, S.I., 1984. 'On Dynamics of Deserts and Climates', pp. 107–120 in J. Houghton, ed., *The Global Climate*. Cambridge: Cambridge University Press.

Rivers, J.P.W., J.F.J. Holt, J.A. Seaman & M.R. Bowden, 1976. 'Lessons for Epidemiology from the Ethiopian Famines', *Annales de la Société Belge de médecine tropicale*, vol. 58, no. 4–5, pp. 345–357.

Sachs, Wolfgang, 1993. 'Global Ecology and the Shadow of "Development"', pp. 3–21 in Wolfgang Sachs, ed., *Global Ecology: A New Arena of Political Conflict*. London: Zed Books.

Sadler, Peter G., 1976. *Regional Development in Ethiopia*. Cardiff: University of Wales.

Said Samatar, 1983. 'Somalia into the 1980s: Problems and Possibilities of Social Transformation'. Paper presented to the Second International Congress of Somali Studies, University of Hamburg.

Sandford, Stephen, 1976. *The Design and Management of Pastoral Development*. London: Overseas Development Institute.

Sandford, Stephen, 1983. *Management of Pastoral Development in the Third World*. New York: Wiley.

Schmitz, Marc, ed., 1992. *Les Conflits Verts*. Brussels: European Institute for Research and Information on Peace and Security (GRIP).

Sen, Amartya K., 1981. *Poverty and Famines*. Oxford: Oxford University Press.

Sharif Harir, 1993. 'Militarization of Conflict, Displacement and Legitimacy of the State: a Case from Dar Fur, Western Sudan', pp. 14–26 in Terje Tvedt, ed., *Conflicts in the Horn of Africa: Human and Ecological Consequences of Warfare*. Uppsala: Uppsala University.

Sherman, Richard, 1980. *Eritrea: The Unfinished Revolution*. New York: Praeger.

Simon, Julian L., 1981. *The Ultimate Resource*. Princeton, NJ: Princeton University Press.

Sobania, Neal, 1988. 'Pastoralist Migration and Colonial Policy: A Case Study from Northern Kenya', pp. 219–240 in Douglas H. Johnson & David M. Anderson, eds, *The Ecology of Survival. Case-studies from Northeast Africa*. London: Lester Crook.

Sobania, Neal, 1990. 'Social Relationships as an Aspect of Property Rights: Northern Kenya in the Pre-Colonial and Colonial Periods', pp. 1–19 in Paul T.W. Baxter & Richard Hogg, eds,

Property, Poverty and People: Changing Rights in Property and the Problem of Pastoral Development. Department of Social Anthropology, Manchester University.

Somalia, 1977. *Rural Development Campaign.* Mogadisho: Ministry of Information and National Guidance.

Somalia, 1979. *Three Year Development Plan, 1979–1981.* Mogadisho: State Planning Commission.

Somalia, 1987. *Five Year National Development Plan. 1987–1991.* Mogadisho: Ministry of National Planning.

Somalia, 1988. *Somali Livestock Statistics, 1987/1988.* Mogadisho: Ministry of Livestock, Forestry & Range.

Stahl, Michael, 1988. 'Environmental Degradation and Political Constraints in Ethiopia', pp. 181–196 in Anders Hjort af Ornäs & M.A. Mohamed Salih, eds, *Ecology and Politics: Environmental Stress and Security in Africa.* Uppsala: Scandinavian Institute of African Studies.

Stevens, Christopher, 1976. *The Soviet Union and Black Africa.* London: Macmillan.

Storas, Frode, 1990. 'Intention or Implication: The Effect of Turkana Social Organization on Ecological Balances', pp. 137–146 in Paul T.W. Baxter & Richard Hogg, eds, *Property, Poverty and People: Changing Rights in Property and the Problem of Pastoral Development.* Manchester University: Department of Social Anthropology.

Sudan, 1982. *Sudan Fertility Survey, 1979.* Khartoum: Department of Statistics.

Sudan, 1990. *Demographic and Health Survey 1989/1990.* Khartoum: Department of Statistics.

Sudan, 1990. *The National Economic Salvation Programme, 1990–1993.* Khartoum: Ministry of Finance and Economic Planning.

Sudan, 1995. *Fourth Population Census of Sudan 1993. Northern States.* Khartoum: Department of Statistics.

Swift, Jeremy, 1977. 'Pastoral Development in Somalia: Herding Cooperatives as a Strategy against Desertification and Famine', pp. 275–305 in Michael M. Glantz, ed., *Desertification.* Boulder, CO: Westview Press.

Tadesse Kidane Mariam, 1985. 'Ethiopia: An Overview of its Priority Environmental Problems, Policy Directives, Objectives, Strategies and Targets'. Paper presented to the International Seminar on Environmental Diplomacy, Ennis, Ireland, November.

Thomas, David & Nicholas Middleton, 1994. *Desertification: Exploding the Myth.* Chichester: Wiley.

Tiffen, Mary, Michael Mortimore & Frances Gichuki, 1994. *More People, Less Erosion – Environmental Recovery in Kenya.* London: Wiley.

Timberlake, Lloyd & Jon Tinker, 1985. 'The Environmental Origins of Political Conflict', *Socialist Review*, vol. 15, no. 6, pp. 57–75.

Touval, Saadia, 1963. *Somali Nationalism: International Politics and the Drive for Unity in the Horn of Africa.* Cambridge, MA: Harvard University Press.

Trevaskis, Gerald, K.N., 1960. *Eritrea: A Colony in Transition.* Oxford: Oxford University Press.

Trewartha, Glen T., 1961. *The Earth's Problem Climates.* Madison, WI: University of Wisconsin Press.

Turner, Billie L., Goran Hyden & Robert W. Kates, 1993. *Population Growth and Agricultural Change in Africa.* Orlando, FL: University of Florida Press.

UNCOD (United Nations Conference on Desertification), 1977. *The United Nations Plan to Combat Desertification.* Nairobi: UNCOD.

UNDP (United Nations Development Programme), 1981. *Somalia: Annual Development Report.* Mogadisho: UNDP.

UNDP (United Nations Development Programme), 1984. *The Nomadic Areas of Ethiopia.* Addis Ababa: Relief and Rehabilitation Commission.

UNDP (United Nations Development Programme), 1990. *Human Development Report.* New York: UNDP.

UNDP (United Nations Development Programme), 1995. *Human Development Report, 1995.* New York: UNDP.

UNEP (United Nations Environment Programme), 1978. *Principles of Conduct in the Field of the Environment for the Guidance of States in the Conservation and Harmonious Utilization of Natural Resources Shared by Two or More States.* Nairobi: Governing Council of UNEP.

UNEP (United Nations Environment Programme), 1982a. *Environmental Problems of the East African Region* (Regional Seas Reports and Studies No. 12). Nairobi: UNEP.

UNEP (United Nations Environment Programme), 1982b. *Marine and Coastal Development in the East African Coast* (Regional Seas Reports and Studies No. 6). Nairobi: UNEP.

UNEP (United Nations Environment Programme), 1989/90. *Environmental Data Report.* Nairobi: UNEP.

UNEP (United Nations Environment Programme), 1992a. *World Atlas of Desertification.* Nairobi: UNEP.

UNEP (United Nations Environment Programme), 1992b. *World Environment, 1972–1992.* Nairobi: UNEP.

UNEP (United Nations Environment Programme), 1993. *Environmental Data Report, 1991–1992.* Nairobi: UNEP.

UNEP (United Nations Environment Programme)/PRIO (International Peace Research Institute, Oslo), 1989. *Environmental Security. A Report Contributing to the Concept of Comprehensive International Security.* Nairobi and Oslo.

UNESCO (United Nations Education, Scientific and Cultural Organization), 1995. *Population and Vital Statistics, mid-year 1994 estimates.* Paris: UNESCO.

UNICEF (United Nations Children Fund), 1996. *The State of the World's Children.* New York: UNICEF.

United Nations, 1950. *Report of the United Nations Commission for Eritrea.* New York: United Nations: General Assembly, Fifth Session.

United Nations, 1995. *Statistical Yearbook*, 40th issue. New York: Statistical Office, UN.

United Nations International Conference on Population and Development, 1995. *Programme of Action.* New York: United Nations.

UN SEPHA (United Nations Special Emergency Programme for the Horn of Africa), 1991. *Situation Report.* September. New York: United Nations.

Wakoson, Elias N., 1984. 'Origins and Development of the Anya-nya Movement 1955–1972', pp. 127–204 in Mohammed Omer Beshir, ed., *Southern Sudan: Regionalism and Religion.* Khartoum: University of Khartoum, Graduate College Publication No. 10.

Wallensteen, Peter & Karin Axell, 1993. 'Armed Conflict at the End of the Cold War, 1989–92', *Journal of Peace Research*, vol. 30, no. 3, August, pp. 331–346.

Waterbury, John, 1987. 'Legal and Institutional Arrangements for Managing Water Resources in the Nile Basin', *Water Resources Development*, vol. 3, no. 2, pp. 92–104.

WCED (World Commission on Environment and Development), 1987. *Our Common Future* (Brundtland Report). Oxford: Oxford University Press.

Westing, Arthur H., ed., 1986. *Global Resources and International Conflict. Environmental Factors in Strategic Policy and Action.* New York: Oxford University Press.

Westing, Arthur H., ed., 1989. *Comprehensive Security for the Baltic: An Environmental Approach.* London: Sage for PRIO/UNEP.

Westing, Arthur H., 1991. 'Environmental Security and its Relation to Ethiopia and the Sudan', *Ambio*, vol. 20, no. 5, pp. 168–171.

Whittington, Dale & Elizabeth McClelland, 1992. 'Opportunities for Regional Development and International Cooperation in the Nile Basin', *Water International*, vol. 17, no. 3, pp. 144–154.

Wistanley, D., 1973. 'Rainfall Patterns and General Atmospheric Circulation', *Nature*, vol. 245, no. 5422, pp. 190–194.

Wolde Amlak Araia, Mohammed Kheir Omer, Adugna Haile & Woldeselassie Ogbazghi, 1994. 'Resource-Base, Food Policies and Food Security: Eritrea Case Study', pp. 84–103 in M.A. Mohamed Salih, ed., *Inducing Food Security*, Uppsala: Scandinavian Institute of African Studies.

World Bank, 1985. *Somalia: Agricultural Sector Review.* Washington, DC: World Bank.

World Bank, 1988. *Poverty and Hunger.* Washington, DC: World Bank.

World Bank, 1990. *World Development Report.* Washington DC: World Bank.

World Bank, 1994. *World Development Report.* Washington, DC: World Bank.

World Bank, 1995. *World Development Report.* Washington, DC: World Bank.

World Resources Institute, 1994. *World Resources 1992–1993.* Washington, DC.

Zeremariam Fre, 1991. 'The Legacy of War', pp. 131–142 in *Greenwar: Environment and Conflict.* London: Panos.

Zewdie Abate, 1991. *Views on Nile Basin Diagnostic Study.* Nairobi: UNEP.

Index